The Gendered Self

The Gendered Self

Further commentary on the transsexual phenomenon

Anne M. Vitale

ISBN: 978-0-557-73533-4

Published by Flyfisher Press,

Point Reyes Station CA, 94956

Also by Anne Vitale PhD

Notes on Gender Role Transtion

http://www.avitale.com

Contents

Introduction

First, a word about the title. If you are familiar with the history of this subject you will be quick to notice that the subtitle refers to Dr. Harry Benjamin's groundbreaking work, *The Transsexual Phenomenon,*[1] published in 1966. Prior to its publication, relatively little had been entered into the medical literature about the subject. What had been published tended to be about cross-dressing (transvestism) in men and gave the impression that the "disorder" was one of perverse sexuality. Benjamin, on the other hand, saw it more as a physical problem whereby one's biological sex (male or female) was incongruent with one's gender identity—one's indelible sense of being male or female. Furthermore, he writes that he believed that gender was indelibly imprinted on the brain either in utero or shortly after birth and that that imprinting could not be changed. Instead of seeing his patients as being delusional and suffering from a sexual perversion, he saw his patients as suffering from a form of intersexuality[2]—having the body of one sex but the gender of the other. It is this paradigm shift in thinking that makes his work so important.

Benjamin wrote his book based on actual case studies of approximately 500 mostly male-to-female patients he saw over a ten year period (1954 -1964). He noted that since psychotherapy was ineffectual in changing how transsexuals felt about their gender identity, transsexuals should instead be treated by hormonal and surgical sex change. Benjamin went on to predict that eventually neuroscience would find that gender identity and expression was hardwired in the brain and not simply a socially constructed aspect of our persona. As we will see in this commentary, that has come to past.

I met my first gender-variant individual in 1976, when the psychiatrist I was seeing at the U.C. San Diego Medical School arranged for me to meet another patient who was dealing with gender issues. With that meeting, a chain of introductions ensued and I was quickly inducted into the greater San Diego "gender community." I met a recently retired nuclear submarine captain, two lawyers, a parole officer, a social worker and four college students. All were in gender-role transition or seriously considering it. Some were going from male to female, others from female to male. Although there were the usual outliers, I was impressed from the very beginning with the intellect and otherwise normal bearing and conduct most of the members of the community expressed.

I was studying for my doctorate in psychology at the time. In my dissertation research on transsexualism I learned that little, if anything, of what I was reading in the professional medical literature about gender issues and transsexualism was in accord with my observations in real life. For some reason, even though Benjamin had opened the door to having an honest discussion about gender issues when he published *The Transsexual Phenomenon,* a surfeit of subsequent authors tended to consider people presenting with gender identity issues as having serious personality disorders. Their most popular choice out of the DSM pantheon was Borderline Personality Disorder, long considered by psychiatrists and psychologists alike as the worst of the personality disorders. These authors often described their clients in peer-reviewed professional journals as being narcissistic, delusional and manipulative. The notion that these individuals were not delusional or that there may really be a biological explanation to what these professionals were seeing never seemed to occur to them. To counter these apparent prejudicial characterizations, I wrote a phenomenological dissertation entitled *History and Resolution of Sex/Gender Integration Needs in Four Male-to-Female Transsexuals.*[3]

Terminology

One of the most vexing problems in writing about gender issues is what words to use. As the transsexual phenomenon settles deeper into the psyche of the every day, new terms arise almost in direct proportion to the growing number of people involved. Those taken to writing about their private experience either dip into countless recombinations of past terminology, or create neologisms out of whole cloth. To make matters worse even agreed upon words such as "transgender" or even "transsexual" seem to have different meanings depending on who is using them. No doubt, the terminology used to refer to gender issues is a lexiconic jumble in search of a convention.

Fortunately most of the new terms are meant to enlighten the non-transsexual world and to help the creators feel as if they fit into a slightly askew but gender normative world. Unfortunately not all words used to describe the transsexual phenomenon are written by people who are supportive of the transsexual's struggle for human dignity. Pejorative words and phrases such as "tranny", she-male, he-she, or even "it" are just a few examples.

One of the intentions I have in writing this book is to bridge the world of the transsexual with the world of the providers dedicated to helping them live full and healthy lives. I am in a unique place in my life to do that. The terminology I have chosen is based on what I have found to be useful in my everyday conversation with other transsexuals, other health care providers specializing in gender issues and in my therapy office helping my gender clients.

There will undoubtedly be those readers who find my use of certain terms and words objectionable. Preferring that I use other terms instead. To them I extend an open hand to argue with me should they be so inclined.

What follows is list of common terms specific to gender identity issues and my intentions when ever I use them in the main body of this book.

Sex and gender. These two words tend to be used interchangeably in society. However in this work they are considered two distinct aspects of what it means to be human. *Sex* is used to refer to the shape of one's genitalia and chromosomal composition (generally, but not exclusively, XX for female and XY for males). *Gender* is used to refer to the hard-wired state of mind associated with one's innate understanding of being male, female, or inconclusive. The gendered self is a root factor in one's identity. As such it plays a leading role in our everyday presentation and interaction with the world around us.

Gender Role. This term is used throughout the book to emphasize the fact that when someone transitions from male to female or vice-versa, they do not actually change their sex nor do they change their gender. Both of those characteristics are an integral physical part of everyone's being and are unchangeable. Transition does, however, alter permanently an individual's gender role. People born male first demasculinize and then feminize to a point where they can no longer be recognized as male or function in the male gender role. Similarly, people born female first defeminize and in turn masculinize to a point where they are unable to be recognized as women and accordingly find it now easier to function in the male gender role. It is this hormonal essentialism that defines one's appearance and subsequently one's gender role.

Transgender: The term "transgender" or "transgendered" has of late emerged as a cover term for anyone who seems to express gender behavior that is counter to what society deems appropriate based on their assigned sex at birth. Transgendered behavior can range from the person who occasionally cross-dresses around the house to individuals who permanently change their gender role through hormonal and/or surgical means. Throughout this book, I refer to this latter group as transsexuals, a term coined by Magnus Hirschfield in 1923. Rather than try to cover the specifics of all the gender variants that are emerging, this book will confine the discourse to transsexualism.

Cisgender: Cisgender or cisgendered is a relatively new adjective coined to provide contrast from the norm to being transgender or transgendered. Essentially it represents the far more common class of people who are comfortable with the gender role assigned to them at birth.

Transsexualism. Transsexualism is a state of existence in which one's sense of gender identity differs markedly from that assigned at birth. As a consequence, transsexual individuals exist, from the very beginning of life, outside the standard male/female binary gendering system. These individuals can more accurately be described as being *gender-variant.* Transsexualism is usually treated by a combination of psychological, hormonal and surgical means. Even though treatment enjoys an exceptionally high success rate in that transsexuals adapt successfully to their new gender role assignment, being gender-variant is as enduring to transsexuals as being unambiguously male or female is to the general population. Counter to what some people believe, transsexualism, if present, is not just confined to just the actual period of physical transition; it colors every aspect of the individual's life from the cradle to the grave.

There are three stages to a transsexual's life. In the first stage, the person is either openly or secretly experiencing generalized feelings of anxiety largely because they are deprived of a sociological outlet for their inner sense of being male or female. Here, I refer to this anxiousness as *gender expression deprivation anxiety.* This period may start as early as age three or four and lasts until it must be addressed in a meaningful manner. The second stage is triggered when the gender expression deprivation anxiety rises to a point where the individual can no longer function productively in his or her daily life and decides to enter treatment. The third stage begins when the individual completes a course of psychological and medical treatment and begins living partly or fully in a new gender role. This gender role transition and its aftermath is covered extensively in this book.

Transspeak-- I have coined the term *Transspeak* to describe the way trans folk commonly use gender and pronoun free linguistic conventions to talk about their pre-transition past without lying yet not exactly telling the truth. Transspeak depends a great deal on the speaker setting up the listener to make the "right' gender specific assumptions while providing plausible deniability should the assumptions ever be questioned.

In transspeak: "When I was a boy" or "When I was a little girl" becomes "When I was a kid" or "When I was a child" Or when referring to one's ex-wife or husband it always gets shortened to "My ex" or "My ex-spouse." Or if still married, just "My spouse". There are countless other examples used daily, especially by the newly transitioned who have little or no recent history in the new gender role to refer to with out revealing their trans status.

There is a serious down side to transsexuals using transspeak. It is a reminder that one is hiding otherwise innocent information from the world regarding themselves that no non-trans person need ever even consider hiding. Used in excess, transspeak can exacerbate the isolation from the norm most transsexuals experience in cohabiting with the rest of humanity.

My Practice

When I opened my psychotherapy practice in 1984, I had no idea my career would lead to specializing in working with people with gender identity issues. Even though I had written my dissertation on the subject and had gone through gender role transition myself five years earlier, it seemed then too rare a condition to even consider earning a living working exclusively with this population. Besides, I was very concerned that my own transsexuality would have a possible adverse influence on how I conducted therapy with this (my) population. Countertransference issues can be very destructive to a vulnerable population.

Despite my initial concerns, and because there seemed to be no one else available, I agreed to start seeing a few indigent clients referred to me by a clinic in San Francisco that had considerable experience in treating sexual minorities but lacked experience in treating individuals with serious gender identity issues.

By 1984, I had been a member of the Harry Benjamin International Gender Dysphoria Association (now known as the World Professional Association for Transgendered Health, WPATH) for five years and had made contact with other therapists interested in gender issues in the San Francisco Bay Area, where I live. Most notable among them were Paul Walker PhD, Lin Fraser EdD, Rebecca Auge PhD, Alice Webb DHS, Luanna Rodgers MFT and Mildred Brown PhD. Together, we formed a peer supervision group that eventually became the Bay Area Gender Associates (BAGA). It is from this base that I began to work in earnest with the gender-variant population.

The role of the therapist in working with individuals struggling with gender identity issues is somewhat unique in the profession of psychotherapy. This is due to the elusive nature of gender identity, the still deeply entrenched social stigma attached to gender variance, and what is often construed by the clients as the "gatekeeper role" of the therapist. (The "gatekeeper role" refers to the client's need to follow the Standards of Care requirement for referral letters from the therapist for hormone therapy and sex reassignment surgery.) Even the very notion of calling a gender issue a "psychological disorder" has been seriously questioned, and there is a strong movement to change that designation in the forthcoming DSM-V.[4,5]

Gender variant individuals are typically voracious readers of all things pertinent to their dilemma. Much of what has historically been written in the medical literature has been by clinicians who have not been friendly to their cause. As a consequence, they are aware of the contentious history between the gender community and some clinicians.[6] Fortunately the animosity between these two groups has eased somewhat over the last ten years, as new authors and a new generation of more enlightened clinicians have emerged. However, one still hears negative echoes of clinicians as self-ascribed "gatekeepers" on some of the gender-related internet forums.

About this Book

I have written this book to share my experience in working with more than 500 gender-variant clients over the last 26 years— covering everything from intake to long-term outcomes. Far from being a medical treatise, this book is about otherwise normal people with a special existential problem that has and will continue to color all aspects of their lives. It has been a privilege to have worked with these outstanding individuals. Despite their almost overwhelming handicap, the vast majority of people I have worked with have shown themselves to be highly intelligent, motivated and courageous. Although it has not been consistent by any measure --there are some notable disastrous rejections to report (see below) --my practice has taught me a great deal about the willingness of many of the people around my clients to give up long-held preconceived notions of who they thought gender-variant individuals were and open up to the new and more accurate life-affirming reality.

On the negative side, I have seen the very notion of gender role transition bring out hatred and spite in once-close families, resulting in abandonment by spouses, divorce and loss of one's children, and loss of long-term friendships, driving the gender dysphoric individual deeper and deeper into despair and far too often to suicide.

A second objective is to pass on to other therapists my philosophy and style of working with gender-variant clients. My approach is simple: I take every individual seriously, no matter how male or female they appear on presentation, how old or young they are, or what social circumstance they are in. I tell each client during our very first session that I have no agenda other than to help them find a way to make their life work. Reaching that goal can vary from finding a minor way to satisfy their need for cross-gender role expression to, if necessary, complete gender role transition. I tell them that I will share everything I know about gender role transition but that ultimately each individual must be ready to accept full responsibility for decisions made along the way.

Finally, this book is about the people who, for no reason of their own making, find themselves intellectually and emotionally separated from living fulfilled lives by the sex of their physical body. It is the story of what it means for thousands of men, women and children who suffer gender expression deprivation anxiety but find a way to go on to live full and successful lives.

Chapter 1

A Brief Description of The Problem

Given gender identity permanency and its obvious importance in the social ordering of one's life, it is reasonable to consider gender identity as essential existential knowledge, knowledge that can not be unknown or separated out from the whole without radically redefining the whole.

Virtually all societies recognize only two sexes—male and female. Nature, on the other hand, appears to distribute the behavioral characteristics normally associated with being male or female along with our gender identity, masculine or feminine in various amounts to each of us. We now know that some people, although assigned one sex at birth based on the shape of their genitalia, (our sex) grow up feeling themselves to be that of the other (our gender identity). How can that be and what can we do about it? In this chapter, I try to answer these questions by referring to recent studies that show evidence supporting the hypothesis that gender identity develops as a result of an interaction between the developing brain and sex hormones and studies that implicate fetal brain development and the complex procedure the embryo goes through in becoming male, female or intersexed and what role the developmental complexity itself plays in creating a potential state of sex/gender incongruence.

The Problem:

In relation to a person's identity, the words *sex* and *gender* have long been used interchangeably in the English language. With the growing knowledge gleaned from gender-variant people, that practice has changed somewhat and we have begun to allot separate definitions to the two words. *Sex* is now restricted to describing the chromosomal, genital or reproductive aspects of the body—male, female or intersex. *Gender* is now being recognized as the word that describes one's sense of maleness or femaleness: traits of masculinity or femininity permanently imprinted in the brain. That recognition of "who we are trait wise" is our gender identity.

Gender may well be the most defining factor in the spectrum of bits and pieces of psyche that comprise the human persona. In fact, gender is so basic to our identity, that most people mistakenly assume our sense of being male or female is defined with absolute certainty by our anatomical sex. Contrary to popular belief, one's indelible sense of being male or female or somewhere in between, and one's chromosomal anatomical sex are two distinct biological elements. Each element develops at different times in different parts of the developing fetus and continues until shortly after birth at which time the window of development closes permanently.

John Money has coined a useful term to describe this phenomenon: Gendermaps.[1,2] Money defines a gendermap as "the entity, template, or schema within the mind and brain unity that codes masculinity and femininity and androgyny." Because gendermap development is highly influenced by hormones produced by both the mother and the developing fetus, sex and gender identification are generally closely matched. But like most aspects of human development, there are no absolutes. As a result, an individual may, as early as the age of three or four, become aware of being caught in the dilemma of being told they have the anatomy of one sex and are, accordingly, a boy or a girl, but being equipped in their brain with a gendermap more typical of an individual of the opposite sex. Thus, although anatomically and socially considered one sex, they feel as though they are the other. It is also apparently possible for an individual to have no clear sense of being specifically male or female. We are more and more coming to believe that gender identity exists along a spectrum rather then an either/or binary system.

Although cross-gender behavior issues are increasingly acknowledged in the popular media, Gender Identity Disorder as it is referred to in the American Psychological Association's Diagnostic and Statistical Manual IV -TR[3] is not a new phenomenon. Indeed, it may be as old as humankind. Reports of cultural anthropologists and others interested in human nature are replete with accounts of cross-gender behaviors that span Classical and Hindu mythology, Western and Asian classical history, the Renaissance, and late nineteenth and early twentieth century studies of preliterate cultures.[4-8]

There is no clearly understood cause for gender variance. However, we have enough information about fetal brain development and the procedure the embryo goes through in becoming either male, female or intersexed, to implicate the complexity of the procedure itself as a cause of the spontaneous sex reversal or potential sex/gender discontinuity.[9] What follows is an abbreviated sample of what we now know about what goes on relative to being gendered physiologically.

Genderizing the Brain:

Evidence of sexual differentiation of the brain has been documented by research. Two studies in particular are of note.

Zhou J.-N, et al.[10] examined the volume of the central subdivision of the bed nucleus of the stria terminalis (BSTc), and found that a female-sized BSTc was found in male-to-female transsexuals. This led them to declare that a female brain structure exists in genetically male transsexuals, supporting the hypothesis that gender identity develops in the fetus as a result of an interaction between the developing brain and sex hormones.

In a follow-up study Kruijver, et al.[11] wanted to know if the reported difference according to gender identity in the central part of the bed nucleus of the stria terminalis (BSTc) was based on a neuronal difference in the BSTc itself or a reflection of a difference in vasoactive intestinal polypeptide innervation from the amygdala.

To do this they looked at 42 subjects to determine the number of somatostatin-expressing neurons in the BSTc relative to sex, sexual orientation, gender identity, and past or present hormonal status. They found that regardless of sexual orientation, men had almost twice as many somatostatin neurons as women. The number of neurons in the BSTc of male-to-female transsexuals was similar to that of the females, while the neuron number of a female-to-male transsexual was found to be in the male range. Hormone treatment or sex hormone level variations in adulthood did not seem to have influenced BSTc neuron numbers. They go on to declare that:

> "Findings of somatostatin neuronal sex differences in the BSTc and its sex reversal in the transsexual brain clearly support the paradigm that in transsexuals sexual differentiation of the brain and genitals may go into opposite directions and point to a neurobiological basis of gender identity disorder."

Even though the brain has both androgen and estrogen receptors, the male brain has been found to be markedly different from the female brain. Not only is the male brain larger and more capable of spatial perception, researchers studying the brains of male and female rats have found evidence that prior to being masculinized, the genetic male brain must first be defeminized,[12-15] a process by which males lose the ability to display female-type behavior.

Once in the fetal brain, testosterone is either metabolized into dihydrotestosterone by an enzyme named 5 alpha reductase or converted to estradiol by an enzyme called aromatase. Counterintuitively, increased estrogen receptor activation is responsible for defeminization while increased androgen receptor activation seems to be responsible for masculinization.[16,17] All this makes clear that there is nothing straightforward about an individual being born with a gender identity that matches their biological sex.

This leads one to consider the possibility that male hormonal surges must occur not only in sufficient amounts in the developing fetus, but must be timed to take advantage of the short time the brain is open to being defeminized/masculinized, forming a predominantly male gendermap. If there is insufficient androgen, or the surge comes too late, the gendermap may be only partially imprinted as male. These disruptions of hormonal surges may come from a variety of sources, including a disorder in the mother's endocrine system such as a hormone-secreting tumor, common maternal stress, medications or some other toxic substance or adverse event yet to be identified.

Being the default condition, genderizing the genetic female brain is far less complex but still subject to having something stray from the norm. If nothing untoward happens, the brain remains female and the individual feels no dis-ease with her body. However, there remains the possibility of a defeminizing/masculinization event to occur for both sexes in utero.

Gender identity, far from being absolute, appears to occur on a continuum, with most people gathered at either end, the rest being somewhere in between. Feelings of discomfort or complete inappropriateness about one's assigned sex do not mean the individual is wrong or ill. It simply means that the assignment made at birth almost universally on the shape of one's genitals can, on occasion, differ from the unseen brain imprint.

Gender Development in Children

Beyond congenital biological determinants, there are at least three well-published theories on gender development in children: biological theory, social learning theory and cognitive-developmental theory. [18,19] The biological theory is based on evidence that high levels of the male hormone testosterone are associated with high levels of aggression in boys and tomboyishness in girls. Social learning theory proposes that gender typing is the result of a combination of observational learning and differential reinforcement. Cogitative-Developmental theory states that gender understanding follows a prescribed timeline, that children recognize that they are either boys or girls by the age of two or three, followed shortly by recognition that gender is stable over time. By the age of six or seven, children understand that gender is also stable across situations.

No matter what theory one chooses to accept, for most children the insight about gender typically goes unnoticed when their sex and gendermap are congruent. However, if there is a sex-gendermap incongruency, the child is left perplexed about his or her gender status and begins a lifelong, often compulsive search for resolution of the discrepancy.

All children naturally comply with the demands of their internal sense of gender. Boys generally express male behavior and girls generally express female behavior, even when raised in closely monitored gender-neutral conditions.[20] If there is any confusion in the child, he or she quickly learns from adults and peers that certain gender-expression behaviors are inappropriate for that individual. This is true even of gender-variant children. Some gender-variant children internalize their dilemma and make heroic efforts to display the gender behavior expected of them, while expressing their internal sense of gender through secret play, cross-dressing and cross-gender fantasies. Others may continue to struggle, insisting that they be allowed to openly express maleness or femaleness irrespective of their assigned sex. Either way, the problem its self becomes subsumed into the child's persona.

The arrival of adolescence compounds the difficulties for people who are gender-variant. Understandably, the development of secondary sex characteristics counter to the young person's hopes and desires is a cause for anxiety and deep disappointment. It is also the beginning of a new determination to resolve "the problem", often becoming the individual's driving force in life. Given the specific social reproach against effeminate men, the situation is especially complicated for gender-variant males. Since the obvious first effort is to accept the physical evidence of their genitalia as the reality of their gender, it is very common to see gender dysphoric males make a special effort to push through these early years of adulthood by engaging in stereotypically macho activities. As life would have it, no outward behavior, no matter how male typical has any influence on internal gender understanding. The net effect of these extreme activities serve only to complicate the individual's social involvement, resulting in even deeper anxiety about expressing true felt gender expression.

This anxiety state is typically characterized by feelings of confusion, shame, guilt and fear. Why do they feel this way when others do not? Why are these feelings so persistent? Even though cross-dressing and cross-gender fantasies provide much-needed temporary relief, these activities often leave the individual profoundly ashamed of what she or he is doing. Closely associated with shame is guilt over being dishonest in hiding their secret needs and desires from family, friends and society. For example, gender dysphoric individuals commonly get married and have children without telling their spouse of their gender dysphoria before making the commitment. Typically, they have the mistaken conviction that participation in marriage and parenting will in itself erase their gender dysphoria. All of this then leads to the fear of being discovered in their lie. Their secrecy and fear have some justification: gender dysphoric individuals are very aware of being called sick, uncaring, selfish and even of being ostracized by the people they love the most.

In summary, Gender Identity Disorder is a real and serious problem. Although we don't know all of the causes of the disease that these individuals feel toward their assigned sex, we can be reasonably certain that it is connected with either a congenital irregularity, an adverse event that occurs in the first few months of childhood or some combination of the two. We also know that every individual's sense of gender, once established, is unchangeable over the individual's lifetime. Men do not suddenly think they are women and women do not suddenly think they are men. This is true for transsexuals as well as those whose sense of gender corresponds rightly to their genitalia. Most transsexuals report being aware of their condition from an early age, between three and seven. The only variable is the individual's ability to tolerate the inherent anxiety of feeling wrongly sexed. If the individual's gender dysphoria is a relatively minor one, cross-gender lifestyle changes in dressing and behaviors may be all that person needs to ease the anxiety. However, if the individual's dysphoria is profound, a simple lifestyle change may be insufficient. In the latter case, the need for more authentic gender expression moves from a lifestyle problem to a life-threatening imperative.

What do we do about the tens of thousands of people—perhaps even hundreds of thousands worldwide—who find themselves in this situation? Fortunately, a course of treatment has emerged over the last half century that appears to be very effective. For those whose gender dysphoria is relatively mild, a low dose of cross-sex hormones taken on an as-needed basis can help. For more severe cases, full gender role transition should be considered. Although seemingly extreme, when done under controlled medical supervision, gender role transition is medically safe and is known to be successful.[21-24]

Chapter 2

What It Means to Be Gender Variant

As a result of the embryonic developmental variances described in Chapter One, the affected individual may be left with between a partial to a full sense of having a cross-sexed gender identity. This chapter describes how the resulting gender permutations affect the individual over his or her life span.

Gender Expression Permutations

Based on hundreds of self reports taken on client intake in the last 26 years, I have noticed that gender variant individuals seem to fall into three relatively distinct categories. For discussion purposes only and with no regard to importance or severity of diagnosis, I have categorized these permutations into Group One, Group Two and Group Three.

Group One (G1) is best described as natal males who have a high degree of cross-sexed gender identity. In these individuals, we can hypothesize that the prenatal androgenization/ defeminization process—if there was any at all—was minimal, leaving the default female identity largely intact. Furthermore, the female expression of identity of these individuals appears very difficult or impossible for them to conceal.

Group Two (G2) is composed of natal females who almost universally report a lifelong history of rejecting female dress conventions along with conventional girls' toys and activities and have a strong distaste for their female secondary sex characteristics. These individuals typically take full advantage of the social permissiveness allowed women in many Western societies to wear their hair short and dress in loose, gender-neutral clothing. These individuals rarely marry, preferring instead to partner with women who may or may not identify as lesbian. Group Two women who do marry tend to choose men who are on the feminine side. Group Two is the opposite gender image of Group One.

Group Three (G3) is composed of individuals who look and act unambiguously as society expects them to but privately identify as the opposite gender. We can hypothesize that the pre-natal androgenization/defeminization process was sufficient enough to establish a workable gender identity that approximated their biological sex but insufficient enough to provide comfort in that gender role. For these female-identified males, and male-identified females, the result is a more complicated and insidious sex/gender discontinuity. From earliest childhood these individuals typically suffer increasingly painful and chronic gender dysphoria. They tend to live secretive lives, often making incrementally stronger attempts to convince themselves and others that they are the gender they were assigned at birth.

As a psychotherapist, I have found female-identified males (G1) to be clinically similar to male-identified females (G2). That is, individuals in both groups have little or no compunction against openly presenting themselves as the other sex. Further, they make little or no effort to engage in what they feel for them would be wrongly gendered social practices (i.e., taking the gender role assigned at birth as the basis of authority). Although some notable exceptions can be found, especially in male-identified females, these individuals—at the time of presentation for treatment—are rarely married or have children, are rarely involved in the corporate or academic culture and are typically involved in the service industry at a blue- or pink-collar level. With little investment in trying to live as their assigned birth sex and with a lot of practice in living as closely as possible to their desired sex, these individuals report relatively low levels of anxiety about their dilemma. For those who decide that physical transition is in their best interest, they accomplish the change with relatively little difficulty, particularly compared to G3 individuals.

The story is very different for Group Three. In the hope of ridding themselves of their dysphoria they tend to invest heavily in sex-typical activities. Being largely heterosexual, they often marry and have children, hold advanced educational degrees and are involved at high levels of corporate and academic cultures. These are the invisible or cloistered gender dysphorics. They develop an aura of deep secrecy based on shame and risk of ridicule and their secret desire to be the opposite sex is protected at all costs. The risk of being found out adds to the psychological and physiological stress they experience.

Transitioning from this deeply entrenched defensive position is very difficult. The irony is that gender dysphoric symptoms appear to worsen in direct proportion to their self-enforced entrenchment in the gendered world they do not identify with. The further an individual gets from believing he or she can ever live as a member of the opposite sex, the more acute and disruptive his or her dysphoria becomes.

Depathologizing Gender Issues

Although the official diagnosis in the DSM IV-TR[1] pathologizes people with gender issues as having a Gender Identity Disorder, I and many others who follow this issue do not think that is an appropriate descriptor. As I have suggested elsewhere,[2,3] the condition would be better described by the term Gender

Expression Deprivation Anxiety Disorder (GEDAD). If left untreated, GEDAD manifests itself differently in each of the five classical developmental stages of life: confusion and rebellion in childhood, false hopes and disappointment in adolescence, hesitant compliance in early adulthood, feelings of self-induced entrapment in middle age, and if still untreated, depression and resignation in old age.

Untreated GEDAD as it is manifested across the five major stages of life.

GEDAD in childhood is characterized by confusion and rebellion

GEDAD in adolescence is characterized by false hopes and disappointment

GEDAD in early adulthood is characterized by hesitant compliance to the norm

GEDAD in middle age is characterized by feelings of self induced entrapment

GEDAD in older adulthood is characterized by depression and resignation.

Childhood

As early as age two and a half, most children begin showing a preference for behaviors and activities consistent with their assigned sex. By age three, they actually refer to themselves as a boy or a girl. Interviews with three-year-olds reveal that they agree with statements such as "Girls like to play with dolls, ask for help and talk more than boys, while boys like to play with cars, build things, and hit other children." [4]

Even the casual observer can see that children place a high priority on gender-appropriate behavior at an early age. Most individuals with gender expression deprivation anxiety report becoming aware that something was not right with their original gender assignment as early as age four. Males describe that, unlike other problems a four-year-old boy may have, discussing wanting to be a girl with friends and family was definitely to be avoided.

Even though the following example dates back forty years, it is safe to say that a boy who wants to be a girl and is willing to admit it can still expect to be "corrected," often in a very stern and firm way:

> Arlene, now in her fifties, reported that in school, at the age of six, she (then he) was forced to stand in front of his first-grade class wearing a large pink ribbon while his classmates were encouraged to laugh at him. He was being "corrected" for having been "caught" playing hopscotch with the girls during recess.

Here is an example of a form of behavioral modification meant to ensure immediate cessation of effeminate behavior in a male.

On the other hand, a girl who wants to be a boy and is willing to admit it can expect far less retribution for her behavior. Girls who affect boyish behavior are generally perceived as cute and the behavior is usually tolerated by friends, family and school officials through childhood. Although they reported mild social pressure to "dress pretty" and be more gentle, none of the male-identified female clients that have presented in my practice shared experiencing behavioral modification efforts like the one endured by the hopscotch-playing boy.

Undoubtedly, there are cases where only guidance and time are needed to correct a gender identity misunderstanding --if a "misunderstanding" is all that it is-- in a child. In others, however, it appears that once gender identity is established, no amount of redirecting can change the child's gender identification. Some boys in particular openly endure the taunts of their peers and castigations of their parents in order to live according to their cross-gender understanding. The Child and Adolescent Gender Identity Clinic of Toronto treats many such children brought in by parents who are concerned over what they believe is unacceptable cross-gender behavior. In reporting on their work at the clinic, Ken Zucker and Susan Bradley report a referral ratio of male children to female children entered for treatment since 1978 (n=249) to be 6.3 to 1.[5] Since there is no evidence that cross-gender behavior occurs more often in boys than it does in girls, a possible interpretation of this statistic is that effeminacy in boys may be considered by parents to be more upsetting and in need of correction than tomboyish behavior in girls.

Given the nature of the condition and the ability of some children to conceal it, it may be possible that most children with gender dysphoria are never diagnosed as such. The undiagnosed children cope by sticking rigorously to the role expected of them. Privately, however, they continue to go deeper and deeper into a highly guarded parallel world of cross-gender envy and fantasy. Given their propensity to be studious, detached and self absorbed, these children often live cloistered lives. These children grow up to form the core of Group Three.

Little is known about gender dysphoric boys who privately struggle to fit into their expected gender role. With no apparent problem, (many adult GID clients report being exceptionally well behaved as children) they simply go unobserved by clinicians studying gender variant behavior. Yet from interviewing adults with gender dysphoria, I can report that the problem, although lacking the current intensity, was as real for them then as it is now.

The underlying feelings most often stated were of detachment and confusion, a sense of not really fitting in, even though family and teachers consistently rewarded them for their artificially affected behavior. One of the most common areas of confusion was the original sex assignment process itself. Although adults may think it simplistic, many children are completely perplexed as to why some children are assigned as boys and others as girls.

Given a tendency toward privacy and modesty in our society, many children, especially those without siblings, often have no way of knowing that there is a physical difference between themselves and those differently assigned.

Andrea, a 35-year-old male-to-female, post-operative transsexual recalls that she was completely perplexed over her assignment as male until at age seven her sister was born. While watching her mother change her sister's diaper, she learned for the first time that her assignment as a boy was based on a real physical difference. Although it cleared up part of the confusion, she realized even at that early age that her identity concerns were far more complicated and serious then she had first thought.

It is common for clients to report thinking in childhood that gender assignment was based on parental preference and therefore open for redress. Girls are especially aggressive in their insistence that they are really boys. Indeed, many are so insistent that they go on to act for all intents and purposes as though they are boys, a pattern they carry into adulthood.

For cloistered gender dysphoric boys it was in the area of peers and activities, especially sports, that the problem was most noticeable. Unable or uninterested in competing in organized boys' activities and having been shuffled decidedly away from playing with the girls, many became reclusive. To add to their confusion, and counter to behavior typically reported in openly gender dysphoric boys, many cloistered boys actually preferred solo play with boys' toys and had little or no interest in girls' toys. For example, I have heard more than one long-time post-op male-to-female transsexual speak fondly of having spent countless hours playing with an Erector Set or a Lionel model train set-up that their father had helped them build. Others described designing and making detailed model airplanes, race cars and sailing ships. The more academic of this group report little or no interest in sports and rough-and-tumble play. To avoid castigation from their peers, they report spending a lot of time reading and studying. Although these children appeared to be normal boys doing what most people would consider normal boy activities, they may very well have been doing so while secretly wearing their mother's or sister's underwear, fantasizing about being a girl.

Like many children faced with difficulties they are powerless to change, such as parental arguments and divorce, gender dysphoric children often seek supernatural help with their special problem. This is usually in the form of praying to God and practicing special religious indulgences. This practice has an inherent opportunity for secondary gain. Almost universally they report that they believed that if God interceded for them by changing their sex, their parents and the world would have to accept them in the new sex without question and thereby exonerate them from what they typically perceive to be a negative and shameful desire.

Adolescence

If there was ever going to be a chance for these individuals to show that they are not really the sex everyone else believes they are, early adolescence is certainly it. Virtually every individual I have interviewed reported wanting desperately to have hidden internal sex organs of the desired sex finally come to life during adolescence and force development of their desired secondary sex characteristics.

G1 boys, who have a strong feminine core identity, typically develop a sexual interest in other boys during adolescence and prefer girls as peer friends. Although they still desire to be girls, they appear to have significantly less anxiety over not being female than that reported by the boys in G3. I believe this is due to the relatively uninhibited open expression of their femininity. For example, Monica, genetically male, was 19 years old when she reported to my office accompanied by her mother. She wore gender-neutral clothing but otherwise presented as female in voice inflection and mannerisms. Monica's mother related that Monica had been more like a girl then a boy all her life. Her parents loved her dearly but thought of her more as a daughter than a son. Over the course of treating Monica, I found that although she was distressed over her male physiology, Monica was otherwise emotionally stable and aware of the seriousness of her situation. Once it became clear that she was her own person and ready to undergo transition, on my referral, she began hormone replacement therapy. With the exception of having to face some extreme religious issues brought up by her much older brother, she accomplished an almost effortless transition from male to female. The presence of a great deal of other family support and little or no investment by the family or Monica in her being male made this transition straightforward.

As sexual maturity advances, the Group Three cloistered gender dysphoric boys often combine excessive masturbation (one individual reported masturbating up to six times a day) with an increase in secret cross-dressing activity to release anxiety. In a post-op group I facilitated, Jenna (age 43) spoke fondly of the delight she experienced as a boy when she would find something of her mom's in the dirty clothes' hamper in the bathroom. Two others in the group laughingly agreed that they too took many a trip to the bathroom for the same reason. At the same time, in their public life, these boys report employing overtly stereotypical efforts to draw attention from their secret desires to be female by affecting appearances of being normally male. This includes dating girls, participating in individual sports activities such as swimming, running, golf, tennis and for some, even body building.

Cloistered gender dysphoric boys appear to others and even to themselves to be heterosexual. Although as a group they are not especially active daters, they clearly prefer to date girls when they do date. Significantly, their dating motives are markedly different from those of other boys. For these boys, being on a date is a chance to spend time with a girl in a way not generally allowed under other circumstances. Dating serves two purposes for these boys. The first is social, as it gives them the all-important appearance of being normal. The second is therapeutic.

Being close to a girl's softness, and even her female smell, has a mitigating effect on gender expression deprivation anxiety. The fantasy is not to make love to her but to actually be her.

The situation is less complex for girls. Having more social freedom in both their dress and behavior codes allows at least a modicum of dysphoric relief. Loose, gender-neutral clothing is typically worn to hide their feminizing bodies and there may or may not be an attempt to appear or act stereotypically female. Many adult female-to-male transsexuals report having adopted a defiant attitude toward the world as a coping strategy. As with all teenagers, gender dysphoric girls must contend with emerging sexuality. Not to be confused with how lesbian girls search out and experience their sexuality with other girls, G3 girls do not consider themselves lesbians *per se* but show interested in other girls in a way that parallels that of heterosexual teenage boys. Those G3 female bodied individuals who do date boys, typically search out boys who are soft spoken and on the feminine side.

Early Adulthood

As more information about the possibility of transition to one's felt gender identity becomes available to the general public, we are seeing genetic males with strong core female identities and genetic females with strong core male identities present to psychotherapy in their early twenties with the clear objective of being sexually reassigned.

The cloistered natal males, on the other hand, typically only start to realize the seriousness of their dilemma at this age. It is common to hear reports of these individuals increasing the intensity with which they try to rid themselves of the ever-increasing gender-related anxiety. Many individuals paradoxically adopt homophobic, transphobic and overtly sexist attitudes in the hope that they will override their desires to be female.

The situation can become so convoluted that some gender dysphoric men come to therapy wanting, almost desperately, to be told that they are not transsexual. That would be understandable if they were simply confused and wanted to get to the bottom of their problem. Unfortunately, their stated preference here appears to be more a form of avoidance of the fear and complexities involved in transitioning than it is an honest desire to remain men. For example, there are natal males who desperately want to have breasts but say they would be terribly embarrassed to have them show in public. There are others who wince at the thought of having a female name like Janice or Mary or Linda. There are also gender dysphoric males who think that the social behaviors that most differentiate women from men are frivolous and unimportant, going so far as to believe that women are "less than" men and being embarrassed about wanting to be like them. Interestingly, these people have no trouble at all with wearing feminine apparel—as long as they can do it in complete privacy.

Perhaps the most insidious form of sexism can be seen in the gender dysphoric male who has attained a respected position in a male-dominated profession. These people routinely assert the common sexist attitude that although women are now allowed a certain professional tolerance, the real players are still men. As more people transition while continuing to work at the same position, these transsexual males see firsthand how public respect between men can quickly turn into private ridicule when a male colleague becomes a woman. Furthermore, some gender dysphoric individuals have confessed to participating in sexist jokes as a way to divert even the remotest suspicion from themselves. Given these seemingly unacceptable obstacles, many gender dysphoric males unconsciously accept certain male-driven notions about women in an effort to purge the need to be female from their mind.

When these individuals are questioned further, it is common to see that they have a deep-seated love-hate relationship with their inner need to be female. While they apparently need do nothing to keep the love side of that dilemma alive, the hate side seems to need constant care and feeding. The danger is obvious: As they see it, if they don't continuously think negatively about women, they might have to face the reality of wanting to be one. In essence, the sexism in this group serves as a cover, providing a convenient and unfortunately, in many social circles, a socially acceptable way to maintain denial.

Another common attempt to "make it" in their birth-assigned gender role by gender dysphoric individuals in this age range is to marry and have children. Unlike their non-dysphoric peers, their attraction toward the idea of family may not be the standard one. Some individuals report telling their partners about their lifelong desires to be the opposite sex before getting married, but the vast majority do not, perhaps from fear of ridicule or rejection, or because they maintain the fantasy that marriage will provide a cure. Many clients report that they were sure that being a husband or wife, as the case may be, would cement in their otherwise elusive sense of being normal. This logic, unfortunately, gets extended to the idea of having children. Although gender dysphoric males are generally no better or worse dads than other men and gender dysphoric women are generally no better or worse at being moms than non-gender dysphoric women, they soon come to realize that what they had hoped would be an answer has instead complicated the possible resolution of their gender issues enormously.

In distinct contrast, genetic females in Group Two who do not seek sex reassignment make little or no concerted effort to be rid of their gender dysphoria. Although they may be deeply disturbed by having acquired female secondary sex characteristics in puberty, many assume an androgynous appearance and affect outright male mannerisms. In larger cities, they may find refuge by taking active roles in the lesbian community and being involved in typically male occupations.

Meanwhile, gender dysphoric people must live in the real world; being subject to the same developmental pressures as their peers. Developmental psychologists refer to the ages between 28 and 33 as a time when individuals reassess their dreams and aspirations. Mistaken interests, family obligations and career demands start to become serious concerns. Women who are reaching the latter part of their childbearing years may have children in school or yet to be born. New decisions have to be made relative to the bulk of life that lies ahead. When someone contending with a gender identity issue reaches this pivotal period, the pressures are magnified far beyond what non-gender dysphoric individuals experience.

Gender dysphoric individuals respond to this critical period in two characteristic ways. Those who have access to information and other financial resources give serious consideration to exploring gender role transition. After an appropriate period of psychotherapy and evaluation by a gender specialist, these individuals almost routinely go on to be physically and legally reassigned to the sex that more closely fits their inner sense of self. Others, who may also be aware of sex reassignment options, may find the idea too impractical or too frightening, deciding instead to entrench themselves deeper into life as a member of their originally assigned sex.

Middle Age

For those who continue to struggle with their gender issues into mid-life, new issues come to the fore. At a time when most people realize that about half of their life has been lived and feel the need to make an accounting of who they are and what they have done, this period can be especially anxiety provoking for the gender-dysphoric individual. Decades of trying to overcome their increasing gender expression deprivation anxiety begin to weigh heavily on the individual. Family and career are now as deeply rooted as they will ever be. The idea of starting over as a member of a different sex seems impossible. The fact that their felt need to transition has increased rather than diminished, despite Herculean efforts, is now undeniable.

These individuals often show up in therapy offices with symptoms mimicking Depersonalization Disorder, Depression or Generalized Anxiety Disorder. They complain of panic attacks, irritability, sleeping disorder, inability to concentrate and recent weight loss. If they are married, there is often serious martial discord due to self-imposed disassociation from the family unit. Job performance may also be affected, involving negative performance reviews, some for the first time, or outright threats of being let go unless they seek help for whatever is bothering them. As the individual is pressed ever deeper into despair, suicidal thoughts may begin to intrude into daily life. Even at this point the individual may be reluctant to discuss their gender issues lest the door be opened to a fear-laden real-world exploration of gender transition. They are consumed by feelings of being inexorably trapped.

John, a 51 year-old genetic male, with a career as a medical research scientist, a 23-year marriage and three children aged 20, 17 and 14, phoned me after experiencing a panic attack severe enough to require emergency attention from paramedics. John gave me only his first name and informed me that I was the first to be told what he was about to tell me. He said he was "gender dysphoric" and that he was "desperate." Feelings that were once "controllable through sheer force of will" had increased to where he was now having protracted periods where he would close his office door, lie in the corner on the floor and weep quietly while curled up in the fetal position, holding his genitals in pain. Other than intrusive and repeated fantasies of being female, he had refused to allow himself any overt form of female gender expression. The only other form of temporary relief came through masturbating. He reported feeling that if he was to cross-dress and be caught, he would dishonor his wife and family. Having attained international recognition for his work, he was also concerned about his professional reputation.

Our work together over the three years was slow. However, with the help of extensive individual, group and family psychotherapy augmented by estrogen replacement therapy, he sought and received the full permission of his family and has taken on a female name and is living full time in the female gender role. She is in the process of renewing and redefining her relationship with her family and has successfully returned to work after an extended leave of absence.

Older Adult

Some gender dysphoric individuals proceed into their senior years with their needs and desires to be the other sex still unresolved. Even now these feelings about the matter may be as strong as ever. The relative freedom of gender expression that women enjoy throughout their lives continues, and there is even less pressure on G2 females to be attractive or feminine than when they were younger. For males, the situation is reversed.

Little is known about these individuals. That they exist, however, is indisputable. Surgeons report performing sex re-assignment surgery on individuals as old as 71. I have personally worked with four gender dysphoric males in their early to mid-sixties and have spoken with colleagues who report working with others in their mid-sixties to early seventies.

The issues these individuals face are now very different. Concerns about how to be a father to young children, maintain a career and establish intimate relationships have lessened. New, less resolvable issues emerge. Along with low self-esteem brought on from years of self-denial, these individuals must now contend with a deteriorating body.

For genetic males, along with balding and paunchiness, there are more serious health issues to consider should an older male wish to transition to the female gender role. Cardiac disorders, gastro-intestinal disorders, diabetes and often liver dysfunction due to a lifetime of alcohol abuse are some of the most common problems they face. Here is a statement Tom, a 63-year-old genetic male made to me upon leaving a "starter" group I facilitate after attending for two months:

"I have recently completed a year and a half of interferon and riboviron treatment for Hepatitis C. That means that anything like hormones could be detrimental to my liver health. No doctor would approve that. Short of that I don't believe that there is any in-between for me given my health, age, appearance, marriage and family. I believe now that I have to live my life as a gentle male and that is most comfortable for me. Not ideal, but most comfortable."

Ironically, a mitigating factor for Tom and other seniors is that the natural aging process decreases testosterone levels in genetic males, resulting in a corresponding increase in estrogen levels. The feminizing effects, albeit mild, are welcomed wholeheartedly. As in hormone replacement therapy for younger men, the natural hormonal changes appear to ease some of the psychological aspects of the dysphoria in seniors. Yet when interviewed, those who chose to remain male speak of a clear longing for what might have been. Senior gender dysphoric individuals—be they genetically male or female—typically report they have been waiting, many since childhood, in the hope that their desire to be the other sex would simply "go away." Like those who are younger, they say in resignation that if they had known the dysphoria was going to remain such a strong force in their lives, they would have braved anything to face their dilemma decades sooner.

Characteristically these people can be described as sad, depressed and deeply resentful. In treating these individuals, the best that can be done is to help them feel better about cross-dressing and encourage them to have contact with other cross-dressers in their age group. Success of sorts can be as simple as helping someone find the courage to shave off a mustache behind which he has been hiding his gender issues for forty years.

Conclusion

Clinically, gender dysphoria shares symptoms often associated with Depersonalization Disorder, Depression and Generalized Anxiety Disorder. Differential diagnosis may be complicated by the client's reluctance to disclose the source of the morbidity for fear of being overcome by real or imagined outcomes of the disclosure.

Gender identity issues can be a lifelong condition for those who find it too difficult to deal with them directly. Each life stage presents new dilemmas and decisions in relation to this core issue. In general, it can be said that the more the individual struggles to rid themselves of gender dysphoria by increasing social and physical investments in their assigned sex, the greater the generalized anxiety and the harder it becomes to restart life sexually reassigned. For those individuals who, despite all obstacles, can transition to a new gender role, it has been shown that gender transition that includes psychotherapy, hormonal therapy and—in most cases—sex reassignment surgery, significantly reduces and eventually eliminates the anxiety entirely.

Chapter 3

Living the Lie

This chapter looks at three of the major components in a transsexual's life prior to treatment. The first (Unlived Lives) deals with the frustration and dissociation essentially defining gender dysphoria. The second and third, guilt and shame, while related to the first, have more to do with cultural expectations and an inability for the transsexual, no matter how morally strong, to live up to those expectations. It is this burden that finally sends the individual into therapy and, for some, gender role transition.

Unlived Lives

"Unlived lives" refers to the frustrations inherent in "waiting": waiting for life to begin. Yes, one is doing what everyone else is doing—going to work every day, watching television, eating in restaurants and so on—but the "doing" has a hollow, spectator feel to it, rather than real time, actual involvement. Life takes on the quality of being in an intolerably long line at the bank or the post office. Everyone seems to get their basic needs met but you.

For most gender dysphoric individuals the waiting starts very early. Little boys wait to be allowed to enter play with the girls as a girl. They wait for a pretty dress to wear or a doll to mother for their birthday. Gender dysphoric little girls manage to get by with tomboyish behavior but they also know they're in a serious struggle for real gender expression.

Unknowingly, with no other recourse, these children start the long wait for the problem to resolve itself. It is common for all children to turn to magical thinking and to expect miracles. Gender dysphoric children are no different. For example, one of the more common appeals these children make to end the intolerable waiting is to pray for God to intervene. Others wish upon a star every evening or make a birthday wish asking for divine intervention when blowing out the candles, then lie about it when asked what they wished for. It's heart breaking to think of all those innocent children at this very moment, praying and wishing for a miracle that in all likelihood will never happen. Fortunately, the plight of gender-variant children is now being taken seriously by a small group of mental health providers.[1,2] It is not divine intervention but it is something of a God-send none the less.

Then there are the adolescent years of waiting. A time that starts full of false hope but ends in further disappointment and confusion. This is a time in life when gender dysphoric boys watch in wonder and heartbreaking envy as girls start to develop into young women. On the other hand, gender dysphoric girls begin a feminization process they wished desperately would stop or not happen at all. As boys endure what they feel is grotesque masculinization, girls endure what they feel is the indignity of feminization. At this point a level of resignation sets in for both sexes, knowing now that the problem is a significant part of their life and may never go away. However, like gender-variant younger children, there is new hope for treatment for adolescents as well. It involves hormonally postponing puberty so that any future decision to go ahead with gender role transition can be made without having to reverse unwanted secondary sex characteristics.[3]

When the gender-variant individual enters his or her early twenties, the efforts to relieve the waiting takes a more practical turn. If a young male can't be a woman, he can at least try to become a man. Perhaps if he joins the armed forces? Perhaps if he gets married? Perhaps if he has children? Perhaps if he starts lifting weights? Perhaps...perhaps... perhaps. If a young female can't evolve naturally into a man, perhaps she can simply continue acting like one. Of course, it isn't the same but it is better than the alternative. However, there are other complications in store for these dysphoric people. Foremost among these is the need for intimacy.

For some gender dysphoric males, being with the right woman eases the tensions of non-participation. Marriage adds social respectability and makes mom and dad happy. The problem is that he almost invariably falls in love with the woman he wants to be and as often as not, he starts living his life through her. This pseudo-life is typically characterized by secretly cross-dressing in her clothes, even going shopping with her and encouraging her to buy clothes that he wants to wear. It usually doesn't take very long for the wife to catch on and the individual to realize that far from providing resolution, marriage has complicated the matter. Now he finds himself waiting to become a wife as well as a woman. Can wanting to become a mother be far behind? Usually not.

In the meantime, gender dysphoric females take full advantage of a world denied their male gender-variant counterparts. Unlike gender dysphoric males, who usually deliberately avoid being associated with gay life, these women often move into the lesbian community. There they have a history of easily finding someone who enjoys their unique blend of maleness and femaleness. They also take full advantage of the social acceptance of short hair on women and wearing gender-neutral clothing. Add a few male clothing items and the image, if not the fact, of maleness is almost complete.

By now the picture is clear. As the decades pass, anxiety and depression place an ever growing burden on the gender variant's life. Typically, life stagnates and becomes something to endure. Some people turn to drugs and alcohol to mollify the pain. When that too becomes intolerable, thoughts of suicide creep in and become the solution of choice.

Guilt & Shame: The Unfortunate Twins

The American Heritage Dictionary, 3rd Edition,[4] defines guilt and shame as follows:

guilt n. 1. The fact of being responsible for the commission of an offense. 2. One that brings dishonor, disgrace, or condemnation. 3. A condition of disgrace or dishonor; ignominy. 4. A great disappointment.

shame n. A painful emotion caused by guilt. 1. The fact of being responsible for the commission of an offense. 2. (Law) Culpability for a crime or lesser breach of regulations that carries a legal penalty. 3. a. Remorseful awareness of having done something wrong. b. Self-reproach for supposed inadequacy or wrongdoing. 4. Guilty conduct; sin. Strong sense of guilt, embarrassment, unworthiness, or disgrace.

Guilt and shame have long been significant components in psychotherapy. This is true no matter what the underlying issue. The reasons can be traced directly to our culture's use of both of these emotions in complex programs to establish moral and social control. Sometimes social manipulation of these emotions are used constructively and sometimes not. Pathology occurs when there is too much or too little of these unfortunate twins.

Excessive guilt and misplaced shame are what therapists typically find when working with individuals struggling with gender identity issues. For example, in my practice, callers setting up an intake appointment often go to great lengths to avoid using such words as cross-dresser, transsexual or transgendered to describe themselves. Fortunately, they give enough clues about why they want to make an appointment without my having to ask them directly. They say they were referred to me by so and so who they know I am seeing for a gender issue. Others will cite having read my web site.

My clients come from all over the country and from all walks of life. They are on the whole upstanding, law abiding, hard working, honest, and productive citizens. As I have noted earlier, typically, they are highly educated, hold medium to high level positions in corporate or government organizations, and are well respected for their abilities. If they have children, they want to be good parents. They appear to be the very model of what society professes to value most in its citizens.

Lying just below the false mantle of respectability, a gender dysphoric's life is often strewn with lies of omission, half-truths, surreptitiousness, broken commitments, and gross manipulation of friends and loved one with the person being painfully aware of the deceptions.

Most gender dysphoric individuals are aware of their off-centered world. Given their condition, which they accurately perceive to be outside the experience of most people, they are often hypersensitive to what society defines as gender-appropriate behavior. In a complicated attempt to conform, most gender dysphorics make dedicated efforts to gain positions of social value. Initially, this is done to convince themselves of their normalcy. When that fails, it is done to convince others of their normalcy. Marrying and having children, especially in gender dysphoric males, is often an extension of these efforts. Unfortunately, even these good intentions eventually become something to feel guilty about.

Many clients come in feeling that they are among the lowest of the low. For decades they have kept within themselves what they believe to be one of the worst secrets imaginable. Long before they give anyone a chance to evaluate their dilemma, they may view themselves as sick, perverted, queer or even an out-and-out freak. They are convinced that if they openly express their inner gender feelings they will be considered uncaring and selfish. Worse yet, they fear, with considerable justification, that they will be ostracized by the people they love the most.

Society's message is clear: stay within the boundaries of behavior allotted to your assigned sex or face possible banishment from all that you know and love. The reasons for this social dictum are both complex and outside the scope of this book. However, I believe it is safe to say that sexism plays a role. In our patriarchal culture, the control mechanism for males is shame. It is expressed through deprecation of all behavior that is not considered masculine. Accordingly, shame expressed by genetic males who wish to be female far exceeds that experienced by females who wish to be male.

The irony for most male-to-female transsexuals is that as males they are forced to participate in institutional sexism from deep within its ugly bowels. All male children learn early that being male is a privileged state. Furthermore, they learn that they are expected to contribute and commit to its continuance. Male-to-female histories often reveal that as boys, no matter how much they envied girls and wanted to be one, still found that being male had its advantages. Unfortunately, boys also learn that to be accepted by their peers and eventually advance from boyhood to manhood, they must denounce behavior that is considered feminine. This forces gender dysphoric boys as young as five or six years old to go underground with their desires to be female. To compensate for this deception, they may make superhuman efforts to at least appear masculine.

Guilt and shame represent deeply ingrained feelings, so ingrained that it is common for each of them to outlast gender role transition. Even though most transmen or transwomen are glad that they are now free of the dysphoria, shame continues to plague some individuals. The most common manifestation of chronic shame is internalized transphobia: a self-loathing and belief that they are sexually perverted. Transphobia is so persistent that on occasion it can be found twenty years after what would otherwise be considered a successful transition. Yet with hard work and a realistic appreciation of the newly assigned sex, even this deep-seated shame can eventually be eased and even turned into pride of accomplishment.

If the transsexual's new life has a sense of authenticity, guilt and shame has a way of easing almost on its own. The key is acceptance, and acceptance comes with time. Parents routinely accept, at least to some degree, their new son or daughter. Similarly, siblings make acceptable accommodations for their new brother or sister. With little or no emotional attachment involved, individuals commonly return to society as responsible doctors, lawyers, parents, spouses, and active business and community leaders.

Chapter 4

Treatment Limits and Options

The medical field has come a long way over the last several decades in learning how to treat transsexualism. We have come to learn that even as much as can be accomplished there remains biological and psychological limits in gender role transition that cannot be transcended. It is an error for individuals undergoing gender role transition and those who are aiding in the process to believe that the process's objective is simply to exchange the individual's polar positions on the male/female binary. It is important therefore that both client and therapist keep a proper perspective on what is medically possible and what is not in order to overcome false expectations. This chapter addresses those limitations.

WPATH and the Standards of Care

WPATH is the acronym for the World Professional Association for Transgender Health. The organization was formerly known as the Harry Benjamin International Gender Dysphoria Association (HBIGDA) established in 1979 at an international conference in

San Diego, California, U.S.A.[1] In 2006, to acknowledge the growing understanding of gender variance and need to recognize a more inclusive population, the name was changed to WPATH. WPATH's mission statement reads in part, "to promote evidence-based care, education, research, advocacy, public policy and respect in transgender health."[2]

At the 1979 meeting the newly minted organization adopted the first version of "Standards of Care for Gender Identity Disorders." In 2001 it released the sixth version,[3] for which the cover page reads in part:

> **The Purpose of the Standards of Care**. The major purpose of the Standards of Care (SOC) is to articulate this international organization's professional consensus about the psychiatric, psychological, medical, and surgical management of gender identity disorders. Professionals may use this document to understand the parameters within which they may offer assistance to those with these conditions. Persons with gender identity disorders, their families, and social institutions may use the SOC to understand the current thinking of professionals. All readers should be aware of the limitations of knowledge in this area and of the hope that some of the clinical uncertainties will be resolved in the future through scientific investigation.
>
> **The Overarching Treatment Goal.** The general goal of psychotherapeutic, endocrine, or surgical therapy for persons with gender identity disorders is lasting personal comfort with the gendered self in order to maximize overall psychological well-being and self-fulfillment.

With the goals in mind, lets first look at the treatment limits.

Treatment Limits:

Five of the most common terms used when working with people struggling with gender issues are *gender dysphoria, Gender Identity Disorder, female-to-male, male-to-female, and sex change*. Because these terms are based on a male/female binary and culturally imposed standards, by definition they are not apt descriptors of people who are gender-variant. If taken literally, they could lead to practitioners misdiagnosing the condition, clients misinterpreting their situation and both client and practitioner having unrealistic expectations for treatment outcome.

The following discussion includes why I find these phrases awkward or inappropriate. Where possible, I offer suggestions for alternative terminology.

GENDER DYSPHORIA-- Before I get into why I have trouble with this term let me confess that common usage has so ingrained the term into the therapeutic lexicon that I still find it useful in talking to others regarding clients who are contenting with gender issues. Although I will continue to use the term through out this book, keep in mind that I think it tends to trivialize a condition that deserves a much stronger term.

Gender Dysphoria is perhaps the most commonly used term to describe a condition in which an individual is experiencing clinically significant levels of gender expression depravation anxiety. I have trouble with both words in the term.

First, everyone I have treated with gender expression deprivation believed that their gender or inner sense of being male or female is simply that, a sense of being male or female. Clients do not complain that their physical sex is "right" and their sense of being male or female is "wrong" so "please fix me." The complaint is that their inner sense of being male or female is inconsistent with their physiology. The idea of changing this inner sense of self, their gendered self, which is such a critical element in the pantheon of elements that compose the essence of who a person is, is anathema to them.

Second, "dysphoria," defined by Marriam-Webster's Collegiate Dictionary as "a state of feeling unwell or unhappy," or in the American College Dictionary as "a state of dissatisfaction, anxiety, restlessness, or fidgeting" is simply too soft a word to describe the deep angst most clinicians see on intake with this population. At best it may be an apt descriptor for individuals who, despite strong evidence to the contrary, are making an extraordinary effort to convince themselves that they are sex/gender congruent.

Typically, at time of presentation these individuals report that either their lives are in ruin, or they are very afraid that if their gender variant condition was to become known they would loose all that they cherish and be ostracized from family, friends and the ability to support themselves. High anxiety and deep depression with concurrent suicide ideation is common. These individuals are not "unhappy" they are miserable.

GENDER IDENTITY DISORDER (GID) -- This term was introduced with the publication of the DSM IV.[4] It replaced the venerable descriptor, "Transsexualism" that first appeared in DSM III [5] seven years earlier.

According to the authors of the *Interim Report of the DSM-IV Subcommittee on Gender Identity Disorders,*[6] the primary reason for creating the new term was to combine the various DSM III-R diagnoses of Gender Identity Disorder of Childhood, Transsexualism, and Gender Identity Disorder of Adolescence or Adulthood, Nontranssexual Type. The authors were also concerned that the term Transsexualism had taken on the singular definition of referring to someone who had decided to transition from one gender role to the other through hormonal and surgical procedures. Clearly a limiting term especially where children were involved. The paper's abstract states that the term Gender Identity Disorder allows "the concept of a spectrum of gender dysphoria rather than discrete levels of symptoms." This was a valuable and insightful contribution to an evolving understanding of what it means to diagnose and treat someone who has a gender issue. However, the time has come for us to use our more recent experience in working with this population to move closer to what clinicians are really treating when a gender-variant individual enters their office.

In the sixteen years that Gender Identity Disorder has been in the DSM lexicon, it has shown to have both aided and harmed the treatment of gender-variant people. On the plus side, the severity of the term has lent a significant level of medical authenticity to the gender-variant condition. Individuals so diagnosed have used it to justify everything from transitioning on the job to keeping their marriages together ("The doctors told me it was the only cure to my illness") to getting insurance coverage for treatment.

On the negative side, referring to a gender-variant condition as a mental disorder has given fuel to those who feel that it should be treated with psychotherapy and not with surgical or hormonal interventions. Another unfortunate unintended consequence of the mental disorder stigma is that it often keeps gender-variant people from seeking help. Thinking that what they are experiencing is a mental disorder (because it has been discussed as such) they strive to privately rid themselves of the problem by distraction (drugs and alcohol) or outright denial.

Although GID may sound descriptive of the problem to those who have their sex and gender neatly congruent, it is often considered offensive to those who have a different life circumstance. Referring to someone's gender consciousness as being "ordered" or "disordered" has, until now, come largely from those living outside the gender-variant experience. After working with more than 500 individuals with sex/gender dimorphic feelings, I have come to believe that if there is any disorder to be treated in these individuals, it is one of understandable anxiety.

There are two prongs to this anxiety. One is a physically induced anxiety that builds over the life span due to a deprivation of estrogenic or androgenic hormones, as the case may be, demanded by the brain but, given the physiologic makeup of their bodies, unable to be produced in adequate amounts. The other prong is represented by the imposition of rigid dress and behavior codes, especially as experienced by pre-treated transsexual males.

In contrast, the treatment plan delineated in the WPATH Standards of Care, Version Six, has it right. The administration of cross-sex hormones after a period of differential diagnosis, educational psychotherapy, encouragement in the exploration of cross-sex behavioral expression and in some cases, surgery, has been found to significantly reduce anxiety, elevate depression and improve the quality of life for a significant number of these individuals.

As was noted in Chapter Two, if we consider the problem as one of being caused and sustained by hormonal and socially ingrained gender expression deprivation, it helps both the therapist and the individual to understand the real issues involved and leads away from the erroneous thinking that one can psychotherapeutically "correct" or "cure" a client's gender identity. A "correction" has yet to be shown to be possible in individuals with a significant level of variance from the more absolute male/female norm.

Considering the transsexual phenomenon as simply a natural variation of gender identity encourages both the therapist and the individual to work toward attaining an anxiety-free resolution that is authentic yet compatible to acceptable gender expression within a binary-based gendering system.

MALE-TO-FEMALE AND FEMALE-TO-MALE

Clinicians and gender-variant people use these terms routinely. Often reduced simply to MTF and FTM, respectively, they have become an accepted shorthand. As handy as the phrases and acronyms have become, they are merely descriptors of a general direction of transition, not a concrete start-to-finish process. As much as clinicians and perhaps even people who present with gender-variant issues may wish, clinicians do not actually take men as most men apparently understand themselves to be and help them become women. Nor do clinicians take women as most women apparently understand themselves to be and help them become men. Gender-role transition is a lifelong process that starts near one gender pole and progresses ever closer without completion toward the other.

Rather than simply changing men into women and women into men, something more profound happens during transition. Post-treatment individuals who are busy going about their daily lives routinely report transition as a being a life-saving experience.

Empirical studies continue to show a significant absence of regret.[7] Conversations with long-term post-operative individuals (25-30 years following surgery) outside the clinical setting show that the "no regret" being referred to in the studies does not necessarily suggest having attained a definitive attachment to one or the other poles of a binary sexing system. Instead, it means that through hormonal and/or surgical intervention these individuals have moved from a place they intuitively knew as wrong to a place that feels far more authentic.

In my opinion, the administration of cross-sex hormones is the key agent in the enterprise. Not only do hormones help the individual attain the desired secondary sex characteristics, they also seem to have a gender-confirming maturation effect on that portion of the brain wherein one's sense of being male or female resides, a place that prier to treatment had been suffering hormonal deprivation.

Beyond hormones, sex-reassignment surgery opens up new possibilities regarding relationships and gender-role social status. The implication is that there is room between the binary poles for significant satisfaction as long as the individual is given appropriate medication, surgery and space to define themselves.

It is common to hear from individuals who have undergone sex-reassignment surgery, some of whom were treated as long as 25 years ago, say they have come to realize that they will always remain in the middle ground of the gender spectrum. Important to them, however, is that this condition continues to be far, far better than their life before starting transition.

SEX CHANGE

Finally, the term *sex change* is also misleading. Technically, sex is defined by chromosomal composition at the time of conception. Obviously, no amount of hormonal manipulation or surgery will ever change that. Combined with the fact that there a spectrum of chromosomal variations possible at conception, resulting in various intersex conditions, using binary sex descriptors as positive markers by society is in reality only relativistic and the idea that one can "change" from one sex to the other absurd.

Although the term "sex change" is rarely if ever used in professional circles, it remains common in the lay press. This indicates a large gulf between what specialists in the field of gender issues understand about sexual assignment/reassignment and what the lay world understands. The term "gender role transition" would be a better choice. I first heard this term used by Rebecca Auge, Ph.D., during a peer supervision group meeting a decade ago. After discussing it in great detail with her, I have since adopted it and encouraged others to do so as well. It is clear that the descriptive language regarding gender issues is evolving. Unlike in the past, the new language does not emphasis or even hint at pathology, disability or disorder. Instead, it is characterized by its inclusiveness and support of individuals searching for and largely attaining a full and healthy life.

The clinical advantages of using less negative and more aptly descriptive terminology not only helps in the psychological recovery of the individual, it also helps those around the client to accept the condition and the client upon his or her reentry into society in the new gender role. Outside the therapy office, therapists must continue to expand their own and society's definition of sex/gender expression beyond the current binary sexing system to be more inclusive of the full range of human gender-variant possibilities.

The Options

In his ground breaking book, The Transsexual Phenomenon[8], Harry Benjamin M.D, concentrating primarily on male-to-female transsexuals, was among the first to note that psychotherapy with the aim of curing transsexualism was ineffectual as a treatment. His remarks are repeated here to show how far back knowledge of the ineffectiveness of trying to change an individual's gender identity through psychotherapy goes. To this day there are no documented case studies showing sustained gender identification reversal due to psychotherapy.

Dr. Benjamin states, "The mind of the transsexual cannot be changed in its false gender orientation. All attempts to do this have failed." He goes on: "Since it is evident that the mind of the transsexual cannot be adjusted to the body, it is logical and justifiable to attempt the opposite, to adjust the body to the mind. If such a thought is rejected, we would be faced with a therapeutic nihilism to which I could never subscribe in view of the experiences with patients who have undoubtedly been salvaged or at least distinctly helped by their conversion."

As mentioned earlier, the major guideline used by most mental and physical health providers for treating gender-variant individuals is the WPATH Standards of Care (SOC). Although the SOC make no attempt to delineate for whom and under what conditions sex reassignment is a viable treatment, it does set minimum treatment schedules and professional qualifications for providers. In addition, WPATH continues to hold biennial conferences at which papers are discussed at length by practitioners from around the world.

Authors who have written books devoted to transsexualism are another source of in-depth medical information regarding treatment options. Here is a partial list presented in order of publication: Green & Money (1969), Kando (1973), Stoller (1975), Koranyi (1980), Lothstein (1983), Walter & Ross (1986) Bolin (1987), Devor (1989), Blanchard & Steiner(1990), Tully (1992), King (1993), Zucker & Bradley(1995)[9]

The topics covered in these volumes include the possible causes of gender dysphoria along with treatment suggestions, including endocrinological and surgical aspects. Although there is still some disagreement about how gender dysphoria begins and who should qualify for hormonal and surgical intervention, there is a remarkable amount of agreement in several important areas. Most psychologists now agree that gender dysphoria qualifies as a subject of clinical attention separate from other disorders. Further, most clinicians agree that the gender identity beliefs people with gender dysphoria hold are profound, deep seated, and non-delusional. Even more significant, outcome studies clearly indicate that when three conditions are met—a proper differential diagnosis, a significantly long trial period of living in the gender of choice, and a satisfactory surgical result— there is only a small incidence of postoperative regret. Indeed, in a review of the outcome literature from 1961 through 1991, Pfafflin and Junge report that fewer than 1 percent of the female-to-male transsexuals who had undergone sex reassignment had any regrets. For male-to-female transsexuals the proportion was slightly higher, though still less than 2 percent.[10] In this study, "satisfaction" was measured by self-report concerning improvement in the individual's psychosocial well being.

There is a rule of thumb that advises every therapist to intercede in a manner that is the most respectful and least invasive to the client. If a minor intervention will do the job then that is the approach the therapist should use. If deeper, more intensive life-changing work is necessary, it should only come after the client has gained insight into how to deal with the ramifications of the potential change.

This axiom is especially important when someone presents with a gender issue. An example of a minor intervention, say for occasional cross dressing, might be simply helping someone understand what they are struggling with and then helping them make allowances for it in their life. On the other hand, if the dysphoria is profound and at a level where the individual is experiencing a clinically significant impairment in one or more important areas of functioning, disability or an important loss of freedom, the level of therapeutic intervention would be correspondingly deeper. In the latter case, sex reassignment becomes a serious consideration.

Before getting into treatment options that have been found to be helpful, let's look at treatments that have been proven *not* to work. I have been asked repeatedly why a cross-dresser or gender dysphoric male isn't given testosterone shots and why a gender dysphoric female isn't given estrogen, on the assumption that more of the hormone that matches their physical characteristics might make them feel more like the sex whose physical characteristics they express. The reason this treatment is not advised is that gender dysphoria is far more complicated than just not enough hormones—it is a problem of the *wrong* hormones. There is some evidence that the male sex hormone testosterone plays a part in establishing male gender identity in newborn males. However, that appears to be about the extent of its gender-influencing ability. Furthermore, people who undergo hormone therapy to bring their felt gender in line with their physical sex are given a thorough baseline physical examination as part of the treatment. Medical records show that, with only rare exceptions, the testosterone levels in males and estrogen levels in women fall within the normal range for their respective sex, so that more of the same is not therapeutic.

The problem is that, unlike non-dysphoric individuals, people with gender dysphoria experience hormone levels that would be otherwise normal (for their sex) as anxiety inducing.[11] It has been shown repeatedly that when testosterone is administered to a gender dysphoric male there is an immediate increase in anxiety. Conversely, when the same individual is given a sufficiently high dose of estrogen (which inhibits the testes from producing testosterone) he typically reports experiencing not only a significant reduction in anxiety but a profound sense of well-being. In a parallel paradigm, gender dysphoric females report feeling a similar sense of well-being when given a sufficiently high dose of testosterone over a long enough period to inhibit estrogenic production in the ovaries.

It is also known that the administration of cross-sex hormones *must* be maintained to sustain the anxiety-reducing effect. It is not unusual for some patients, feeling better after starting hormones, to believe they are cured and no longer need to continue the medication, so they stop taking it. The predictable result is an almost immediate return of their gender dysphoria. If there is any physical test to determine who should seriously consider partial or full transition, taking cross-sex hormones is it.

Conclusions:

With undeniable physical evidence for gender dysphoria and with options being limited by nature itself, the task of the provider must center on helping the individual make the best adjustment possible to a situation that no matter the effort put into it will never provide a perfect outcome. Given the fact that all of this is happening in an adverse social environment it is the provider's duty to educate those in his or her care not only of what is medically possible but the limitations as well.

Chapter 5

Transition: Therapeutic Interventions

Chapter 5 deals with the role of the psychotherapist when presented with a client experiencing gender identity expression deprivation. A lot has changed for the better over the last 26 years I have been practicing. Twenty five years ago it was common for someone presenting to a therapist with a gender issue to be considered psychotic and treated as such. In recent years, a more contemporary, depathologized model has emerged, requiring the therapist to be more of a case manager, educator and guide rather than a adjudicator of mental illness. Much of what is presented here is philosophical in nature. I believe that it is not the therapist's place to tell someone what sex they should be or what gender role would best suit them. Instead, it is the therapist's job to help the client find a way to make their life work. If gender role transition is the best answer than so be it, but keep in mind that it is only one of the possibilities and will not work for all individuals presenting with gender expression issues.

Diagnosis

It is commonly understood that each of us has a gender understanding of self that is more or less masculine or feminine. As we have seen earlier, since gender and biological sex develop at different times and in different places in the body during embryonic development, gender identity may on occasion have little or nothing in common with the individual's biological sex.

For most individuals, small variations in absolute gender identity add human interest to their personality. Typically, these individuals experience no psychological or biological imperative to change their physical appearance. However, for individuals who experience a major dichotomy between their sex and their inner gendered self, the disparity can be crippling. Without treatment, it is common for these individuals to live their entire lives in a chronic state of longing to become sex/gender congruent. In clinical terms this is known as a chronic state of dysphoria. If the dysphoria is severe enough, individuals can become suicidal and/or so chronically depressed as to be unable to function.

Given the apparent unalterable, imprinted nature of gender identity, the therapist is limited to helping the individual learn to live with his or her condition. In actual practice, depending on the level of anxiety and the desires of the individual, treatment can range from recommending occasional cross-gender expression (guilt-free cross-dressing) to making a referral to a physician for moderate cross-sex hormone therapy. In more severe cases, it may be appropriate to first educate and then start the individual on a program that includes complete hormonal and surgical sex reassignment: gender-role transition.

For many therapists, the hardest part of treatment is diagnosis. It is common for gender dysphoria to exist along with other, more obvious psychological difficulties. Generalized anxiety, mood disorders such as chronic depression (dysthymia), depersonalization disorder and/or substance abuse to ease the pain are four of the more common comorbidities. The challenge is discerning which disorder is primal and what disorder(s) are secondary.

What gender specialists are looking for varies from the complex to the very subtle. The more challenging cases involve multiple personalities (dissociative disorder) in an individual. There may be a co-existing highly developed female self and a highly developed male self, with both personalities fighting for dominance. Equally difficult to treat are people who find it difficult to hold on to any sense of identity for an extended period of time. These people have often been abused as children and some may believe that becoming a member of the opposite sex may be an answer to the trauma they continue to experience.

More subtle cases involve cross-dressers exploring the difference between occasional cross-dressing needs and the possibility of being transsexual. In addition, effeminate gay males have been known to present to check out the possibility of changing their sex with the belief that if a little effeminacy is fun, more might be better.

Of Concern

Everyone agrees it would be great if we had an evidence based psychometric test we could give clients to confirm or deny a GID diagnosis. Perhaps some day one may be developed but for now we will have to continue to rely on very careful judgement and peer review of difficult cases.

Benjamin's Sex Orientation Scale.--
Despite its unproven accuracy, one of the most common metrics therapist (and some clients use to self diagnose) is the Benjamin Sex Orientation Scale. [1] Established in 1966 it has come to obfuscate the diagnostic picture more than it has helped. The Benjamin SOC is problematic in two ways.

First, by establishing a scale that runs from "transvestites" to "true transsexuals" Benjamin unintentionally created a hierarchical scale where by some therapist believe only "true transsexuals" should be cleared for gender role transition. In fact, rarely if ever does a therapist have such individuals report for intake. Most gender clients fall far short of that ideal but certainly deserve having the therapist's keeping an open mind as to what may be accomplished.

Second, the Benjamin Sex Orientation Scale is essentially a self-report form that assumes that the individual is not inhibited in talking about his or her dysphoria. I have noticed, however, that most gender-variant people keep their situation a secret for decades prior to seeking help, and disclosure is usually very difficult. For example, I am no longer surprised to hear that I am the first or second person (after their spouse) they have ever told about their dysphoria. Some people report getting physically ill when they disclose their situation to someone for the first time. I have found that the more I put my clients at ease and get them to open up about their true feelings of gender expression deprivation, the higher up the Benjamin Scale they seem to rise. Therapists who misconstrue the client's inability to talk about their gender issues as a sign of there not really being a true gender identity problem risk doing great harm to their clients.

Gender Identity Clinics--On the whole, gender identity clinics have provided a great service to individuals struggling with gender issues. Without them I doubt if the legitimacy of the condition would have been established and sex reassignment given the credibility it now enjoys. There are, however, serious drawbacks inherent in the clinic setting when treating gender issues. First, this kind of formal mental health clinical model creates a doctor-patient relationship that tends to over-pathologize the problem. Pathologizing, by its very nature, puts all of the power of the relationship in the hands of the clinicians. In the clinic setting, the doctor is well and the patient is ill. To further distance the individual (and the individual clinicians) from the process, committees sit periodically and review the current case load. Essentially, clinicians, some of whom may not have ever met the client, decide if the individual is a candidate for hormone therapy and, eventually, if the patient is eligible for sex reassignment surgery.

I find this autocratic approach to the treatment of gender issues problematic. In actual practice, the all-power-to-the-doctor format often creates an adversarial relationship between the clinicians and those they are trying to help. Professional responsibility is, of course, important, but there are ways of exercising therapeutic responsibility that encourage the client to take on the lion's share of the overall responsibility. In my experience, the more responsibility the client is given in resolving his or her gender issues, the more responsibility they take and the more successful the outcome.

Another negative result of the committee approach is that it tends to seek out straightforward cases and weed out any that may require special attention. The net effect is to eliminate from treatment not only individuals who clearly do not belong in the program, but also people who want to explore just how much of a gender problem they have before making the commitment the team may be looking for. Extended, deep inner work can easily be misinterpreted by a team to mean that the individual is too uncertain and should be eliminated from the program. Other individuals who are often left out include people dealing with the social ramifications of being a cross-dresser, people who do not want to complete a classic transition, or post-op people who are still learning how to adjust to the new role in life.

Case Management Model

A better model is the classic, one-on-one approach of psychotherapy combined with case management and consultation for difficult cases. In such a model, everyone who feels they have a problem is admitted into the office and there are no limits or expectations placed on the therapeutic process. Assuming that the client has the psychological resources for making intelligent decisions, it has been shown repeatedly that they can successfully decide for themselves what level of gender transformation, if any, they are most comfortable with.

I believe it is not the place of the therapist to define their client's gender identity. As in working with any therapeutic issue, the therapist's place is to help the client come to terms with the implications of whatever an insightful therapeutic search reveals.

If after careful evaluation, gender role transition advances to being a serious consideration, the therapist should provide and monitor a structured program that carefully integrates hormonal and psychotherapeutic treatments. The therapist should be ready to support the client by educating the individual on what to expect, both psychologically and physiologically, prior to and as transition progresses. Support should include just "being there" on an as-needed basis and aiding the client in addressing the inevitable issues that arise with family members, employers, and friends.

Conclusions

In sum, people hire a gender therapist because they are overwhelmed by anxiety due to obsessive cross-gender thoughts. As in more routine therapy, treatment options should responsibly follow the level of discomfort of the individual. Whatever the course of treatment, in order to relieve the anxiety it must be focused on helping the individual evaluate and then accept their internal sense of gender. Fortunately, gender expression needs are becoming better understood and much can be done to help gender dysphoric individuals short of sex reassignment. Because a gender identity crisis often comes in the prime of life, sex reassignment means coping with a series of difficult trade-offs. It's a little like surviving a natural disaster. Sex reassignment not only means getting used to physical changes, it also requires picking up the pieces and starting over again, rebuilding a life. The therapist should be willing and ready to provide ongoing help with that phase of the process as well.

Chapter 6

Transition--Clinical Interventions

The Triadic Treatment Plan (psychotherapy, hormonal therapy and surgery) of the WPATH Standards of Care serves as the armature around which the mechanics of gender-role transition are built. The first few sessions with a psychotherapist, being the most crucial, are discussed in detail. The need to walk through the procedure, its limits and options prior to making more serious physical commitments is discussed. For serious cases, the administration of cross-sex hormones, ancillary procedures such as chest reconstructive surgery for female-to-males, hair removal and facial feminization surgery for male-to-females, and the variations and options for sex reassignment surgery for both types are examined.

The Standards of Care

The WPATH Standards of Care (SOC), as discussed in Chapter Four, was first developed in 1979 and has been updated five times since. It is a living document, with international recognition, that has served as the foundation for the high percentage of successful outcomes in individuals with gender-variant issues. Originally the SOC was written as a guide to help therapists help clients transition from one gender-role to the other. Although it is excellent in that regard, it is not much help in working with individuals who have no desire to transition but still need to find some resolution to their gender expression deprivation anxiety. The forthcoming, Version 7 of the SOC will in all likelihood correct this shortcoming.

The SOC outlines a Triadic Treatment Sequence.[1] Like most treatment statements, it reads as a cold and unfeeling document. To make it useful requires a well-tempered and seasoned therapist to humanize the endeavor. The triadic sequence, as stated in the SOC is presented here, followed by my commentary on each of the stages.

The Triadic Sequence

Stage I: Evaluation, diagnosis, individual/group psychotherapy, education.

Stage II: If applicable, referral to physician for hormone replacement therapy. Also used for confirmation or rejection of diagnosis. Continued individual/group psychotherapy.

Stage III: Monitor gender role transition, serve as an authoritative intermediary in matters pertaining to work and family. If applicable, make referral for gender reassignment surgery after a minimum of one year of living full-time in the preferred gender role. Post-op follow-up as necessary.

Stage I: Evaluation, Preliminary Diagnosis, Psychotherapy and Education (Minimum of 12 sessions)

Intake session. The intake session may be the most significant session a gender specialist has with their client. It either opens the door and sets the stage for all the work that follows, or it may be a brief, one-hour interchange where both parties decide that the "fit" is not right. In the latter case a referral to one or more other therapists is required and that may be the last the therapist sees of the individual. Interestingly, no one has ever said they did not want to work with me when asked directly at the end of the first session. However, a significant number of potential clients were never heard from again or canceled the second session and then dropped out of sight. Normally, that would be just the nature of getting any therapeutic relationship started. Remember, however, that gender-role transition is scary stuff; the fact that the individual returns for the second visit is important and should be used as a positive therapeutic opening.

Although the routine varies from therapist to therapist, all prospective clients can expect the intake appointment to be somewhat structured. The reason for the structure is that along with learning about the presenting problem, there is a certain amount of legal, ethical, and financial business for the therapist to attend to. If the therapist is to take on the responsibilities of being the client's mental health provider, she needs to know as much about the prospective client as possible in that first hour. Furthermore, she has to do this while showing the client that she has the willingness, expertise, capacity and above all the empathy to help resolve the problem.

The advent of the Internet has changed much about how I acquire and accept new clients. Today it is common for most therapists to have an educational web site allowing the prospective client a have neutral space in which to do a preliminary evaluation of the therapist prior to asking for the first appointment.

A primary feature of my web site is my New Client Information page.[2] Here I offer notice of my adherence to the WPATH, SOC, my contact information, hourly fee, insurance policy and payment schedules, including that I accept credit cards. I also require prospective clients to send me, via email, a short autobiographical statement that centers on how their gender issues are affecting their current living situation. Much about the prospective client can be learned by such a statement. I not only pay attention to the content, I pay attention to thought organization, length of the of discourse and phraseology.

No matter what the presenting problem, the therapist should provide a safe place for the individual to work. Even so, only a very small portion of what is really going on with the client will come out in the first session. An experienced therapist should be able to make a good estimate of the extent and kind of psychotherapeutic work that will be required. Some therapists give a formal mental status exam, but I chose to do so informally, phrasing my questions as a normal part of a therapeutic discourse, flagging only answers of concern to revisit. I urge therapists who encounter a potential client with a gender issue that seems far more than they feel qualified to work with to refer to someone with experience in gender identity issues. The issues involved in working with people who have gender issues are complex and often border on what some therapists might feel to be unnatural.

Most new client inquiries come via my website. By the potential client agreeing to the conditions I have delineated there, I know ahead of time if the primary reason for contacting me is to work on a gender issue. I ask all interested parties to use the word INQUIRY in the subject line of the email they send. For queries about therapy that don't come through the Internet, I ask directly how the person got my name. We are into coded territory here and it may make all the difference between having the client feel comfortable or not. Like me, most therapists have clients who have totally different issues, so it is important not to assume that all new callers are calling regarding a gender issue.

No matter what the circumstances that led the individual to be in my office for the first time, I have found it very helpful to be as warm and informal as possible during that first meeting. This is often a very frightening moment for many gender clients. For example, my office is on the second floor at the top of a long flight of steps that run up the outside of the building to a small landing outside the door of my waiting room. In one of my groups, a relatively new client was relating to the others how, at the first meeting, he stood at the base of those stairs for several moments in anticipation of the unknown that awaited him at the top. He had to struggle with an overwhelming urge to return to his car and drive off. A second participant in the group smiled and related that they too had a similar experience on first confronting "the stairway" that eventually led them upward, literally and figuratively, into a whole new life.

Over the two and half decades I have worked with gender-variant individuals, I have attained an easy, matter-of-fact relationship with the subject of gender dysphoria and I frame my inquiries to reflect an everyday familiarity with gender variance and cross-dressing issues. I have come to believe that this reassures the individual in two ways. First, the individual can immediately tell that she or he is not going to be taken for some sort of sexual pervert or, even worse, considered crazy and sent off to a hospital for electroshock therapy. Second, if I do my job well, the individual will start to gain a sense of confidence in me at a critical stage of the therapeutic relationship. Obviously, confidence in the abilities of the therapist is critical in all therapeutic relationships, but when dealing with gender identity issues, it can literally mean the difference between life and death for extremely distressed individuals—a situation that is tragically far too common.

As the intake hour nears its end, if I feel that the client and I are not a therapeutic match, I tell the individual outright, giving them my reasons for not being able to work with them and at least one or two referrals. However, if I feel that I can work with the individual I gradually begin to volunteer more information about my professional qualifications, my therapeutic orientation and my qualifications as a gender specialist. This also includes a short description of what I expect from my clients. I let the individual know that I take a case management approach to my gender work and describe my professional association with other mental and medical health professionals who eventually may become involved in the therapeutic process. I also make it clear that I have no agenda for them other than to help them make their life work. (I know I have repeated that last sentence but I do so because of its importance.) I reiterate that I have the ability, qualifications and more than enough professional associations to see them through gender-role transition should our work lead to that. If the client agrees to work with me, the intake session is closed with the discussion and signing of several standard disclosure forms. I require each person to read and sign a form describing the Limits of Confidentiality and a form describing my general business practices. A third statement, if the situation calls for it, consists of my treatment guidelines.

With business matters completed and agreement to continue working together in place, it remains only to set a regular appointment schedule. My hope is that the next time the client and I meet, the situation will be far more relaxed for both of us and the real work of resolving the client's gender issues can begin.

Second session: Preliminary exploration of presenting problem. If the client is showing symptoms of being gender dysphoric, I consider Gender Identity Disorder (GID) 302.85 per DSM IV as the working diagnosis. I evaluate for co-morbidity and check to see if the client is self-medicating (taking cross-sex hormones obtained on the street or over the Internet). If the client is self-medicating or reports abuse of controlled substances, I make an immediate referral to a physician for a physical evaluation. If the client is married, in an intimate relationship or has children, I let the client know that I am open to seeing anyone they would like to invite to a session or two. I also make referrals to a marriage and family therapist knowledgeable about gender identity issues in cases where that is called for.

Follow up sessions: Objectives and Progress Evaluation. Many adults with gender identity issues find effective ways of living that do not involve all the components of the triadic treatment sequence. Although some individuals manage to do this on their own, psychotherapy can be very helpful in bringing about the discovery and maturational processes that enable gender identity comfort.

Psychotherapy often provides education about a range of options not previously seriously considered by the patient. It emphasizes the need to set realistic life goals for work and relationships and seek to define and alleviate the client's conflicts that may have undermined an otherwise stable lifestyle.

The establishment of a reliable, trusting relationship with the patient is the first step toward successful work as a mental health professional. This is usually accomplished by competent, nonjudgmental exploration of the gender issues with the client during the initial diagnostic evaluation. Other issues may be better dealt with later, after the person feels that the clinician is truly interested in and understands their gender identity concerns—the issue they came to work with in the first place.

It almost goes without saying, but it's important to keep in mind that the goal of a gender specialist should be to help the person live more comfortably within a gender identity of their own understanding. I firmly believe only the client can make that determination. Given the physiological and sociological obstacles involved, especially in male-to-female transitions, this may be a difficult task, requiring weeks, months or in some cases even years to sort out. Even when these initial goals are attained, mental health professionals should discuss the likelihood that no educational, psychotherapeutic, medical, or surgical therapy can permanently eradicate all vestiges of the person's original sex assignment and previous gender role experience.

Psychotherapy is a series of verbal exchanges between a therapist who is knowledgeable about how people suffer emotionally and a client who is experiencing distress. Typically, psychotherapy consists of regularly held 50-minutes sessions. The psychotherapy sessions initiate a developmental process that strives to enable the client's history to be appreciated, current dilemmas to be understood, and unrealistic ideas and maladaptive behaviors to be identified. Under no circumstance should the therapist attempt to "cure" or "eradicate" what may be diagnosed as a gender identity disorder. Instead, the goal should be to establish a long-term stable lifestyle with realistic chances for success in relationships, education and career within a frame work of the client's self-perceived gender identity and expression.

Gender-role transition by its very nature requires a collaborative effort by multiple practitioners: psychotherapists, speech pathologist, electrologist, physicians, and specialty surgeons. The primary practitioner in this list is the psychotherapist. The therapist must be certain that the patient understands the concepts of eligibility and readiness toward moving from one step to the next. Too restricting an adherence to rules (i.e. the SOC) can cause a stalemate between a therapist who seems needlessly withholding of a recommendation and a patient who seems too profoundly distrusting to freely share thoughts, feelings, events and past negative relationships.

Options for Gender Adaptation. The activities and processes that are listed in the next section have, in various combinations, helped people to find more personal comfort. These adaptations may evolve spontaneously during psychotherapy. Starting off slowly in integrating cross-sex gender-role expression in daily life may be all that needs to be done; however, the client may ultimately elect to pursue hormonal therapy, the real-life experience, or genital surgery in the future. There are some who say encouraging minor cross-sex behavior leads one down a slippery slope. That may indeed be true, but in my experience, individuals in treatment tend to take matters of gender expression only as far as they feel comfortable.

Action items...

Biological Males: Cross-dressing unobtrusively with undergarments, in a unisex manner or in an outright feminine fashion for protracted periods. Changing the body through hair removal by laser, electrolysis or body waxing. Having minor plastic cosmetic surgical procedures (small breast implants for example). Developing grooming, wardrobe, and vocal expression skills. Referrals to cross-dressing clubs, support groups and organizations.

Biological Females: Cross-dressing unobtrusively with undergarments, in unisex clothes, or in an outright masculine fashion. Changing the body through breast reduction surgery, breast binding (though not recommended), weight lifting, applying theatrical facial hair; padding underpants or wearing a penile prosthesis (packing).

Both Genders: Learning about transgender phenomena from support groups, communication with peers via the Internet. Becoming familiar with relevant lay and professional literature about legal rights pertaining to work, relationships, and public cross-dressing. Involvement in recreational activities of the desired gender; episodic cross-gender living.

Processes: No matter what the direction of full or partial transition, the individual requires the acceptance of personal homosexual or bisexual fantasies and behaviors (orientation) as distinct from gender identity and gender-role aspirations. It is important that the individual accept the continued need to maintain a job, provide for the emotional needs of their children, honor spousal commitments. It important for the client to identify triggers for increased cross-gender yearnings and effectively attend to them. This typically involves developing better self-protective, self-assertive, and vocational skills to advance at work and resolve interpersonal struggles to strengthen key relationships.

Stage II: Hormone Replacement Therapy

Prior to making a referral for hormone replacement therapy (HRT), it is incumbent on the therapist to make certain that the client knows as much as possible the effects HRT will have on them. Most adult clients have a clear understanding of the obvious physical changes that will ensue but seem to have little to no knowledge of the way cross-sex hormones permanently change brain size and brain function and expand emotional behavior.[3,-7]

It is well documented that cross-sex hormones have a mitigating effect on patients suffering from severe gender dysphoria.[8] The effect is so marked, in fact, that the treatment is used to confirm or reject the GID diagnosis. Fortunately, psychological outcomes precede permanent physiological secondary sex characteristic changes, making it an ideal tool for diagnostic confirmation or contraindication. Some practitioners feel uncomfortable with administering cross-sex hormones to anyone, so it is important that a referral be made to a "gender friendly" physician who is well versed in the administration and monitoring of patients taking these hormones.

Effect of HRT on Adults

The effects of hormones differs from patient to patient. It usually takes at least two years for hormones to reach their maximum physical effect,[9] and taking more than the recommended dosage neither increases the speed or the effectiveness of the drug.

Genetic males treated with estrogen and an anti-androgen (usually spironolactone in the USA) will experience breast growth (size increase will be limited by family genetics), mild redistribution of body fat, marked decrease in upper body strength, softening of the skin, a loss or thinning of body hair and a slowing or stopping of male-pattern balding, should that have been in process.

Within weeks to a month, the average genetic male on a feminizing hormone regimen will notice a decreased ability to gain and maintain an erection. Upon ejaculating, which often becomes painful, the patient will first notice the semen losing its milky color—the testes stop producing sperm—eventually becoming completely clear and reducing in quantity to a clear drop or two as the prostate gland slowly goes dormant. With the exception of breast development, most male-to-female physical changes are reversible once a person stops taking feminizing hormones and as long as the testes remain intact.

Genetic females treated with androgens (testosterone) can expect even more dramatic changes, both physical and psychological, from HRT. In contrast to genetic males taking estrogen, some masculinizing changes of genetic females taking androgens are not reversible. For that reason, special attention must be given in educating genetic females about to start androgens to make certain they understand the implications of what they are about to do.

Among the permanent changes, female-to-male transsexuals can expect a deepening of the voice, clitoral enlargement, minor atrophy of the breast, increase in body hair and even the possibility of male-pattern baldness in time. Among the reversible changes, the individual will experience an increase in upper body strength, weight gain, a marked increase in sexual arousal and a decrease in hip fat.

A little-known fact about both sexes migrating to living in the opposite gender role is that no matter what the direction of transition, transitioners seem to undergo a "second adolescence." There is a marked retro shift in behavior from the adult they chronologically are toward temporarily acting more like a teenage boy of a teenage girl.

This second adolescence seems more marked for male-to-female individuals. Like cisgendered teenage girls, MTFs become very self-centered, are overly concerned with their appearance and often spend hours on the phone or online talking to others going through a similar experience. Spouses staying with their partners through the process complain about this frustrating aspect of the transition more than any other. Behavior befitting a teenage girl does not seem to sit well with a significant other when practiced by a 40- or 50-year-old partner.

For genetic females migrating toward living in the male gender role, there is also the tendency to be self-centered, but unlike their MTF sisters, FTMs tend to be overly aggressive in response to everyday situations, become a bit clumsy as their body enlarges and some times have trouble controlling their new testosterone-driven libido. Another problem is that they often forget that they no longer have female privileges when it comes to interacting with men. For example, they often stand too close in conversations with other men and expect courtesies men do not commonly extend to each other, such as insisting that they pass first through a doorway or deferring to them when ordering in a restaurant. FTMs may also be surprised by the sexually explicit way men often speak about a particularly attractive woman or women in general when they think they are in exclusively male company.

There are also limits to hormone therapy that should be made clear to the patient. For example, along with the obvious limits of estrogen not being able to completely erase entrenched secondary masculine features, such as beard growth, although the testes go into stasis and actually atrophy to a degree, they remain able to resume producing androgens should the exogenous estrogen levels be reduced to a point where they are no longer tipping the endocrine balance toward female.

I have known transwomen who have stopped taking estrogen long enough for their sperm count to rise high enough in order to bank sperm for possible future use. One such woman, after years of being post-op, decided she wanted to have a child, withdrew her banked sperm, hired a surrogate to carry her child and is now the mother of an 8-year-old girl. Mom and daughter are doing just fine, thank you.

The same scenario should be explained to someone considering transitioning to the male gender role. Although introduced exogenous androgens masculinize the female body and force the ovaries to go into stasis, the body remains able to resume producing estrogen at levels where ovulation could return if androgen levels have been reduced significantly.

There are several reports of individuals who have long since transitioned to living in the male gender role temporarily stopping taking androgens in order to become pregnant.

Later Stages: There seems to be a natural progression in gender-role transition. As the hormones slowly change one's body and as more of the important people in their lives have been notified of the transition plans, the individual generally starts to accept his or her status as someone who is "in transition."

If the impetus is to go forward, the therapist should gradually start encouraging and supporting new gender-role behaviors. For example, by now the client should have given serious thought to a new name befitting the gender role to which they are aspiring. Use of the new name and appropriate pronouns should be incorporated into the therapeutic framework, not only to help the client get used to the terms but also to see if they fit comfortably. Given the Western cultural bias against effeminacy in males, this can be especially hard for genetic males learning how to live in the female gender role.

Life in one's assigned birth sex may not have been comfortable but it is all the client has known. If transition is to go well, it is imperative that the individual find a release from the hold of the old gender role. Although it is not generally thought of, male-to-female transsexuals not only evolve toward becoming more womanly, they must also stop holding on to the identity they have been counting on for survival for decades—their entrenched old male identity. Essentially, she must stop being a man. That is what the SOC means when it says that a client who is presenting for surgery should show "Demonstrable progress in consolidating one's gender identity"[10] prior to being referred for sex reassignment surgery.

Stage III Sex Reassignment Surgery:

Sex reassignment is not "experimental," "investigational," "elective," "cosmetic," or optional in any meaningful sense. It constitutes very effective and appropriate treatment for transsexualism or profound GID.[11]

The WPATH SOC requires every candidate for surgical procedures (with the exception of cosmetic procedures such as facial feminization surgery for males and in some cases, bilateral mastectomy for females, where the size of the breast would make it impossible for the individual to succeed in living in the male gender role) to have a minimum of one year of living full time in the new gender role before being approved for sex reassignment surgery. Often referred to as the real life experience (RLE), this requirement is an important one. Despite the RLE being considered by some individuals in the online gender community as an artificial gatekeeper restriction imposed on their life by the medical profession, it is really an opportunity for the client to prove to themselves and others in their life that they can do this and to reinforce that it is serious business. A responsible surgical candidate holds to the absolute restrictions, knowing that to falter might indicate unreadiness, while success can serve to consolidate their new gender status within themselves.

The real-life experience tests the person's resolve, the capacity to function in the preferred gender, and the adequacy of social, economic, and psychological supports. It assists both the patient and the mental health professional in their judgments about how to proceed.[12]

To aid in this important shift, the therapist should be ready and willing to provide logistical support as needed. The laws regarding the legal registration of gender identity vary from state to state in the U.S and they vary from nation to nation outside the U.S. It is incumbent on the therapist to be aware of what is legally possible and what is not. The therapist should be ready to write letters and sign forms to aid in the client changing his or her driver's license, Social Security status and other identification papers. The therapist should consider it part of their duty to educate family, friends and their client's employer should the client make such a request.

Facial Feminization Surgery (FFS)

Pioneered by Dr. Douglas Ousterhout in the 1980s and 1990s, facial feminization surgery (FFS) has fast become one of the most sought after surgical procedures by transwomen. It has now become common for sex reassignment surgery to be put off for years in order to pay for all or as many of the incremental feminizing features as one can afford. For good reason, individuals transitioning to female feel that having a face that is unambiguously feminine helps them integrate more easily and quickly into society as women.

Among the most requested procedures are hairline advancement, forehead contouring, brow lift, rhinoplasty, cheek implants, lip lift, lip filling, chin re-contouring, jaw re-contouring and Adam's apple reduction. Although FFS has its limitations, especially with profoundly masculine face structures, in general the results are amazingly successful. The procedures are expensive, often limiting the number of procedures a person chooses.

There are no letters of referral required for FFS but most surgeons prefer that the patient be well advanced in beard removal and to have been on estrogen long enough for the hormones to have taken their limited but natural face feminizing effect.

Voice surgery for Transwomen

Surgery to raise the pitch of the voice has not yet proven to be effective. Results to date have left patients with a higher-pitched voice but one that sounds unnatural; what some people call a Minnie Mouse voice. Even if surgery is ever perfected for voice tone and pitch, it can do nothing to invoke female voice inflections. That would still have to be a learned skill. For now, voice surgery is not recommended and deemed to be less effective than voice feminization training provided by a speech therapist. With practice, determined individuals are capable of achieving a voice that may be of a lower pitch but otherwise indistinguishable from that of cisgendered women.

Beard removal for Transwomen.

Of all the sophisticated medical procedures available for the transwoman, getting rid of facial hair remains one of the greatest challenges. The process is simple enough; however, it is expensive, painful and can take up to several years for the individual to be beard free. There are two ways to kill and remove beard growth. One is by electrolysis the other is by laser treatment.

With electrolysis, a trained operator, using a fine needle electrode, inserts the probe into each hair follicle while applying an electric current sufficiently high to kill the follicle. The effectiveness or "kill rate" depends on the skill of the operator and pain threshold of the patient. Each pulse is described as feeling like a hot bee sting. Re-growth is said to appear in about 50 percent of the treated hairs, so the process must be repeated over a span of several months until the hair no longer reappears. There are thousands of hair follicles in the average male beard, making electrolysis a long, expensive and painful process. Skin or hair coloration are not a factor in electrolysis. In the right hands, the procedure is safe and sure.

Laser hair removal uses a laser beam designed specifically for hair appellation. This is a relatively new way to rid individuals of facial hair. It is a medical procedure that uses an intense, pulsating beam of light that targets dark pigment in the hair. The result is an intense heat that works down the shaft of the hair to destroy the follicle. In most cases, laser hair removal slows hair re-growth, but it takes several treatments to provide an extended "hair-free" period.

Unlike electrolysis, laser treatments are faster in that they treat a patch of beard hairs in one pulse rather than treat one hair at a time. Operators generally "clear" a whole section of the face in one sitting. It is also reported to be slightly less painful than electrolysis, often likened to having someone snap a rubber band on your face. The drawbacks, however, are significant. Best results occur when the patient has a dark beard and a light complexion. Further more, laser may be totally ineffective on white or blond beard hair. There is yet another, more important concern, no laser clinic claims to achieve 100 percent permanent hair destruction. Re-growth may occur and they often recommend maintenance sessions.

Orchiectomy:

This procedure, more commonly known as castration, is usually employed in cases where the individual either can't afford a complete sex reassignment procedure and has been on a high dose of estrogen for a period that approaches "too long," or the physician prescribing the hormones feels that the high dose of estrogens needed to feminize the individual is contraindicated for some other health reason. Without the testes, the estrogen dosage can be reduced to a mere fraction of what it would otherwise take to reduce male secondary sex characteristics and induce feminization.

This is a simple outpatient procedure wherein both testes are removed from the scrotal sack through a narrow opening. The slits are placed so that scarring will not interfere with possible future use of the scrotal sack in creating the labia in sex reassignment surgery. Most surgeons require at least one letter from a therapist before performing the surgery.

Genital Reconstruction Surgery--Male-to-Female

Genital reconstructive surgery (GRS) is the terminology used by the SOC to describe the final, irreversible surgical process in gender-role transition. It is sometimes referred to as gender reassignment surgery or sex reassignment surgery. What it should never be called is "sex change" surgery. Whatever we call it, we know that society places a strong emphasis on the shape of one's genitals to identify one's gender, and no doubt about it, this procedure is all about changing the shape and function of one's genitals. It has taken many years of exploration and the skill of dozens of surgeons to master the procedures, but we are now very good at altering male genitalia to look and function as female genitalia.

Procedures may include orchiectomy (removal of the testes), penectomy (removal of the penis), vaginoplasty (creation of a vagina), clitoroplasty (creation of a clitoris), and labiaplasty (creation of major and minor labia). Techniques include penile skin inversion (the most common procedure), pedicled rectosigmoid transplant (where the surgeon takes a portion of the patient's colon to use as a moist lining for the neovagina), or in some cases—usually in repairing botched procedures or to add more depth to the vagina—a free split skin graft to line the neovagina. Sexual sensation as well as appearance is strived for and often obtained.

Care after surgery involves maintaining the vaginal vault. — Typically, four days after surgery, the packing is removed from the neovagina and the transwoman begins a lifelong responsibility for keeping her neovagina open. With the wounds not yet healed, the new post-op begins a regimen of regular dilation procedures. This involves the insertion of a series of progressively larger stents into the neovagina and keeping them in place for a total dilation time of half an hour. Some surgeons suggest repeating the process four times a day for the first month, then to slowly diminish dilation periods over time. The final recommendation, six or so months post-op, is for dilation once a day.

Chest Reconstructive surgery (Top surgery)

Female bodied individuals who are experiencing sever gender dysphoria and are considering transitioning to the male gender role, often must contend with large breasts; breasts too large to bind to give the allusion of maleness. Although the SOC states that FTM individuals can undergo bilateral mastectomy at the same time they start hormone replacement therapy, many surgeons will do the surgery on the strength of a letter from a gender specialist verifying the diagnosis of GID and the intent of the individual to live permanently in the male gender role.

There are several different surgical approaches to chest reconstruction surgery. The surgeon is confronted with individuals with breast that are barely there to breast that are outlandishly large. In addition, in many cases the surgeon must deal with disfigurement due to years of breast binding.

DESCRIPTION OF CHEST
RECONSTRUCTIVE PROCEDURES as
described by Michael Brownstein MD[13]
"In a 2 to 3 hour operation, the surgeon performs a bilateral mastectomy and reconstruction of the nipple and areola. It is generally performed under general anesthesia on an outpatient basis.

The nipple/areolar reconstruction uses grafts of existing nipple and areolar tissues to construct a symmetrical, properly positioned and sized male nipple/areolar segment.

The patient with very small breasts may be a good candidate for the "keyhole" or subcutaneous mastectomy accomplished through a small periareolar incision. The advantage of this procedure is the smaller incision and the lack of need for grafts. A good result requires residual skin to contract naturally so as to leave no excess of skin.

A letter of referral from a therapist may be required, indicating the appropriateness of the procedure and timing of this stage of your transition. "

Genital Reconstruction Surgery--Female-to-Male

Surgical procedures available for individuals migrating from living as women to living as men do not enjoy the same level of success as those for their transwomen sisters. Creating a highly functional neophallus surgically has proven to be a daunting task. The current procedures require removing a muscle from the forearm, upper leg or belly, rolling it into a penile shape lined with mucosa from the vagina, rerouting the urethra and then transplanting the whole apparatus to a position in the genital area. The process requires multiple surgeries, has a relatively normal appearance but needs to be artificially "erected" for intercourse. The procedure has other functional problems, is very expensive and may after all of that, still end in necrosis and will need to be removed.

Given the cost and physical limitations of phaloplasty, most FTMs opt for a far less invasive clitoral release procedure called metoidioplasty.[14,15] It is a one-stage procedure that involves lengthening and straightening the enlarged clitoris to create a micro-neophallus. The urethra is lengthened and directed through the micropenis to enable voiding while standing. The procedure is often accompanied by scrotoplasty, which is the creation of a scrotal sack from enlarging the labia and the placement there in of testicular prostheses. Other procedures may include hysterectomy (removal of the uterus), salpingo-oophorectomy (removal of the ovaries}, and vaginectomy (closing of the vagina).

Who is qualified to perform the surgery?

The Standards of Care declare:[16]

> The surgeon should be a urologist, gynecologist, plastic surgeon or general surgeon, and Board Certified as such by a nationally known and reputable association. The surgeon should have specialized competence in genital reconstructive techniques as indicated by documented supervised training with a more experienced surgeon. Even experienced surgeons in this field must be willing to have their therapeutic skills reviewed by their peers. Surgeons should attend professional meetings where new techniques are presented.[17]

Clearly, the number of surgeons worldwide qualified to perform sex reassignment surgery is limited and most people have to travel thousands of miles to be treated. On the whole the cost for male-to-female genital reconstructive surgery is less expensive than the far more complicated genital reconstruction surgery involved in the multiple female-to-male procedures, especially for citizens in countries without a national health care service. Even in countries that do have a national health service, the number of surgeries allowed each year are rationed. It is not uncommon for a significant number of individuals to face the fact that the surgery may not ever be performed no matter how strong their desire. In addition, insurance coverage for these procedures is all but absent.

Although sex reassignment surgery is not required for someone to live permanently in the gender-role opposite that assigned at birth, many wish to have it after satisfying the SOC requirement of living full time in the new gender role for at least one year. At this point in treatment, the patient should be well educated about the choices of surgical procedures available and specialty surgeons of record. Final approval for SRS requires two letters of referral from mental health professionals qualified to treat individuals with GID. One letter must be from the patient's primary therapist. The second letter provides a second opinion as to the readiness of the patient to undergo surgery. A list of therapists qualified to give a second opinion should be provided by the primary therapist.

Chapter 7

The Reality of the Real Life Experience

For all its effectiveness and life saving properties, transition delivers the individual into a life state where membership in a binary gendered world will remain forever elusive. Altered permanently are one's relationships with one's parents, siblings, spouse, children, friends and ultimately with oneself. This chapter looks at the plusses and minuses, the gains and losses and what transitioning individuals come to understand they are becoming. This is generally known as the pre-op or Real Life Experience (RLE) period.

The RLE is sometimes referred to as the Real Life Test (RLT). In fact a "test" of the client's ability to survive the one to two year period living full time in the opposite gender role prior to qualifying for SRS was the original intent in earlier versions of the SOC. Although it wasn't stated directly as such, the requirement was there more or less to protect the clinicians from legal and professional discipline should the transitioning individual sue for malpractice.

The RLE still provides a level of clinical legal and professional safety, but most clinicians now view the RLE as a client based reason for it being in the SOC. This paradigm shift puts the responsibility for success or failure on to the transitioning individual where it belongs. Only the client knows if they are being honest with themselves in their ability to handle full gender role transition. Any cheating (temporarily reverting back to the old gender role due to inconvenience) is in fact harm they are doing against themselves not the clinicians. In my view most transitioning individuals know this intuitively and take the RLE on as a major personal challenge. A responsible transitioner thinks of the RLE as less of a test but as an opportunity to move their persistent gender needs out of the shadows and out into the real world for all to see. What happens from that point on is, of course, unknown. At the minimum the RLE will tell them with certainty the rightness or wrongness of their actions.

Transition is Hard...Very hard!

Gender-role transition changes everything. Other than the trauma of being forcibly ejected naked from the womb and into this world, there is no greater all-pervasive life experience a human being can have. Transition is hard—very hard. The process either opens the door to a new and prosperous life (albeit with significant limits and qualifications) or it can lead to absolute misery. This is not a game to be played lightly. That it is played at all is due to the empirical proof that medically induced gender-role transition qualitatively improves the lives of gender dysphoric individuals.[14]

Relationships

Relationships are both the bane and the hope for people dealing with gender identity issues. Recognition by parents, a spouse, one's children, significant others and friends as the gender they feel themselves to be is extremely important to gender dysphoric individuals. Ultimately, it is the need to relate honestly with the rest of humanity that forces people to admit first to themselves and then to others that they have a gender identity issue. That all should go well upon making that revelation, is the hope. The fear is the realization of possibly losing friends and family and becoming absolutely alone in life. As it turns out, we have come to know that something in the middle is the most likely outcome.

Those who do best during and after treatment realistically accept the condition for its inherent difficulties and not as they might have envisioned it in their fantasies. That includes learning that maintaining the essence of old relationships is all but impossible.

Transitioners must make a drastic readjustment in every relationship they have ever had. As sad as that may sound, it is not necessarily a bad thing. Often, relationships with friends and family improve as they take on a more authentic quality. Time and the unknowable effects transition will have on others are the key factors in adjusting old relationships and establishing new ones.

Not surprisingly, the effect of opening to others is directly related to the extent one has to go to resolve the issues within oneself. If resolution comes through disclosure of being a cross-dresser, the impact will probably be localized to the immediate family and perhaps a few close friends. If, on the other hand, resolution includes taking on a completely new identity by transitioning from one gender role to the other, the effect on everyone involved in the individual's life will be profound.

For genetic males who from time to time take on a temporary female persona, relationship issues revolve around incorporating that activity into what would otherwise be considered a normal family lifestyle. In this situation a lot depends on whether or not the spouse or significant other was aware of their partner's cross-dressing before committing to the relationship. Although there is no guarantee, the spouse or significant other is far more likely to try to accommodate and include their partner's cross-dressing activity as a favorable feature if disclosure comes early in the relationship than if it is announced several years and two kids into a marriage.

If the relationship is heterosexual, the non-cross-dressing partner will naturally wonder how all of this affects their long-held sense of their own sexual orientation. If the notion of being gay or lesbian does not sit well with the partner of a gender dysphoric male, there could be an extremely negative reaction to the situation. Moreover, children, and especially teenagers, themselves in the throes of discovering sexuality, often have problems with the situation. It is not unusual for teenagers to find their parents an embarrassment under normal circumstances; having a cross-dressing or transitioning parent complicates matters immensely.

Although there are husbands who contend with cross-dressing/cross-living wives, wives who initially accepted cross-dressing are more apt to complain that the spouse seems to cross-dress more than what she believes is healthy for the relationship. Wives often report that their spouse's cross-dressing has become a self-indulgence that takes attention away from the marriage. Family or couples therapy usually helps work out these problems.

When resolution for the gender dysphoric individual calls for a complete and permanent change of physical appearance, that level of resolution has a far more profound effect on relationships—for both genetic females who hormonally and/or surgically transition to live and function as men and genetic males who hormonally and/or surgically transition to live and function as women.

The key here is appearance. No matter how supportive the friend or partner may be while transition is still in the talking stage, the first sign of physical change almost invariably forces a reassessment of that support. It may not necessarily mean a complete loss of the relationship, but as the transition progresses, the interchange between the two parties typically undergoes a radical redefinition. I have noticed that the more intimate the relationship before transition, the more likely the relationship will be radically changed. In contrast, more casual relationships, if they don't simply drop away from disinterest on behalf of all concerned, often take on more meaning as formerly hidden common interests are discovered.

Interestingly, the first attempt people entering transition make at maintaining old relationships is to reassure family and friends that only their appearance will change. Explaining that the new person will look different but still be the same person they have always known. They don't realize that this is simply not true. Gender-role transition far exceeds a simple change of appearance. In fact, if that is all transition did, it probably wouldn't be worth the trouble. Upon starting to take hormones, people routinely report a profound improvement in their sense of well-being and the rightness of their action. This occurs long before they or anyone else becomes aware of significant physical changes. Clearly, something far more profound is occurring. As their sense of self changes, their needs and self expression in relation to others will change as well.

Another element that drastically affects relationships is that gender role transition is an inherently selfish act. It has to be. No society I am aware of schedules in gender role transition as part of the natural development of life. Without an equivalent permissive space now allocated to childhood and adolescence and even old age, the transitioning individual --going through a remarkably similar experience to a second adolescence --has no other choice but to selfishly impose his or her own self serving interest on friends, family and community. To not do so invariably leads to failure. Some individuals find being self centered morally wrong and impossible to get beyond. These individuals either suffer interminable delays in transitioning (i.e. prolonging the agony for all involved) or worse yet give up and revert back to a state of chronic gender dysphoria, a situation that can and often leads to suicide.

It may be an impossible expectation to put on friends, family and community at this point in the short history of gender role transition but until further understanding of the biological necessity of gender role transition is common place and our concept of what is morally correct when issues of gender authenticity are involved changes, all parties will continue to suffer a needless and artificially imposed obstacle to life affirming gender role transition for those in need.

One of the more profound and usually unexpected events is that the individual's sexual preferences may change. A genetic male who is used to being in a sexual relationship with women may realize a strong, new, attraction to males once he, now she, is free to both look and act feminine. The new feelings may be triggered by the sexual response a new presentation to the world evokes in others or it may be due to the effects of the estrogen on the brain. The actual cause is unknown. However, when and if it happens, it is common for the individual to say that it is too new and too interesting to ignore. Apparently, a relationship unlike anything the individual has experienced in their old life can be upon them without their doing a thing. The individual may find themselves acting a bit awkward at first (not unlike a teenager experiencing his or her first love), but after a few mistakes and a little advice from others, they usually get the hang of it. I know of three of these unexpected relationships that have gone on to become long-term heterosexual marriages. One of the marriages is in its eighteenth year, a second in its fourteenth year and the third just celebrated its tenth anniversary.

There are many other variations on intimate relationships available to transsexuals. Three of the most common are male-to-female individuals who are now in lesbian relationships and female-to-male individuals who are now in gay relationships. It is also common to find two transsexuals who have met each other during transition to be very much in love and living together in what appears to everyone else as a same-sex couple. There is yet another, less obvious relationship model to discuss. In this case the individual chooses to live alone and interact with a world of friends and family of origin. As one might expect, these individuals are usually older transsexuals who have had their share of intimate relationships in the past and are now totally involved in their professional and community concerns. I know of artists, psychologists, musicians, professors and writers who are quite content with this way of living their lives.

Child/Parent Relationships

Transparent and child relationship especially for those still in transition, varies from family to family. Early on, transsexualism was commonly conflated with homosexuality or even sexual perversion. In keeping with the times, it was commonly believed that children had to be protected from this bizarre and little understood phenomenon lest it confuse the child in a critical period of psychosexual development. Essentially, it was believed, falsely, that transsexualism was a learned lifestyle behavior and that children would be especially vulnerable to corruption should they be exposed to it.

We know a lot more now about how children respond to a parent's transitioning. More couples then ever are opting to stay in the marital relationship and help the children adapt to the new paradigm of having a transitioning or completely transitioned parent. In 1998, Richard Green published a study entitled *Transsexuals' Children*[5] that looked at 18 children from nine families—including ten boys and eight girls. There were six male-to-female parents and three female-to-male parents. The children's age range was 5 to 16 years. Interviews focused on two areas that were often thought of as being potentially problematic for the children: their own gender identity and peer group stigma.

Dr. Green found that none of the children interviewed could fit the DSM IV criteria for gender identity disorder and that there was no clinically significant cross-gender behavior in the children. He also found that three of the children were selective in "informing peers of the transsexual status of their parent, informing only those whom they thought they could trust with the information and who would not tease or spread it indiscriminately. Three children experienced some teasing; it was transient and resolved. The remainder report no problems."

Significantly, the interviews show that the children were far more concerned about losing the love of the parent or that something might go wrong in the process rather than what the parent looked like. For example, when a 14-year-old daughter of a female-to-male transsexual was asked if she cared if her mother transitioned to the male gender role, she replied "I said, no, as long a you are the same person inside and still love me. I don't care what you are on the outside…Its like a chocolate bar, It's got a new wrapper but it's the same chocolate inside."

My own experience bears out the truth of this study. Increasingly couples in my practice are opting to at least try to stay together after one of them transitions and I am seeing first-hand how well children are adapting to the new gender role of a parent. The key seems to be how well the non-transitioning parent seems to be dealing with the transition. As long as the transition appears to be free of conflict (at least in front of the children) the children routinely cooperate.

Children under five seem to have the least difficulty with the situation, probably because, even though children typically identify with being a boy or a girl prior to age three, they don't really understand gender permanency until about the age of seven.[6] Five- to ten-year-olds are a little more perplexed by it all, but have relatively little trouble as long as they believe that their parent will continue to love them and provide the security they had learned to expect. Predictably, teenagers can be expected to be a bit more anxious about it all. Most of the anxiety comes from fear of peer rejection.

Although there are some notable exceptions, the adult children of transsexuals seem to have the most difficulty with their parent's transition. Perhaps it is because a transitioning parent disrupts the status quo in a manner too extreme to be easily handled as a young adult. Concerns I have heard in my office include: What will I tell any future partner? How can I have children with no grandparent to give them? Is there a chance that my children will be transgendered now that it is in the family? Unfortunately, some adult children of transitioned parents have opted out from even acknowledging their transsexual parent or, at the least, minimizing contact. This is especially painful around the holidays and can exacerbate difficulties in the underlying marital relationship.

Of note is the record of a 49 year old male-to-female client I recently worked with who was in a long term marriage. The couple had two young sons and one grown daughter. Once it was agreed that the couple would make a serious effort to keep the family together specific effort was made by the client to remain in his, now her, children's life. The client continued to take her 10-year-old son to a nearby park to play ball—often being complemented by other parents for being one of the few moms who played ball with their son. The client also went on to attend her older daughter's college graduation and met the parents of her daughter's fiancé and eventually attended the daughter's wedding. The older son gave the bride away.

Work Place Issues

Gender-role transition as it pertains to the work-related experience has changed dramatically over the last thirty years. In the late 1970s and early 1980s there was little or no attempt made by transitioning individuals to stay in the same job position. It was almost taken for granted that the transitioning individual would be dismissed outright if they should attempt to hold their job. The record is rife with stories of teachers, doctors, lawyers, engineers, store clerks, and even truck drivers being summarily fired upon announcing their intention to transition. The only exception seems to have been people in civil service positions who were protected by progressive dismissal laws. A probation officer I knew who transitioned from the male-to-female gender role in 1978, reported that although her experience was far from pleasant, especially in the first few years, she stayed on successfully in her job with the county until she retired twenty-five years later.

Although far from universal, laws protecting transsexuals' rights to work in the gender of their choice have proliferated in the last several years. Whether it is a direct result of mandated civil rights laws at the local and state level or simply the result of good common sense, there has been a steady improvement in the attitude employers have toward their transgendered employees. The Corporate Equality Index (CEI) put out annually by the Human Rights Campaign Foundation (HRC) clearly shows how far we have come and how fast progress has been made. Every year since 2002, the HRC foundation has measured non-discrimination and diversity training policies in *Fortune* magazine's 1,000 largest publicly traded businesses, *Forbes* magazine's 200 largest privately owned firms and *American Lawyer* magazine's top 200 revenue-grossing law firms. In 2008 the HRC reported:

> 1195 employers achieved the top rating of 100 percent [in 2008], compared to 138 employers that received perfect ratings in the previous year. Collectively, these businesses employ 8,318,000 full-time U.S. workers. When the Human Rights Campaign Foundation Corporate Equality Index was launched in 2002, only 13 companies received scores of 100 percent. Fifteen of this year's top ratings were achieved by employers appearing in the CEI for the first time.[7]

Many transitioning individuals now find the way clear to stay on the job, removing a strong impediment to the whole process. Unfortunately, professional sports programs and military organizations worldwide remain steadfast against retaining or admitting transsexuals into their ranks.

As more people stay at their old position upon completing transition, unanticipated issues have come to light. There is a major difference between being identified by your coworkers as either a man or a woman and being identified as a transsexual.

Friends who were once loose and friendly with their chatter and camaraderie may now go out of their way to avoid contact. Some even cite religious or moral issues. Others may simply be overwhelmed by the new social situation and feel that it has been unfairly inflicted upon them. Then again others who seem to have been distant in the past might come forward to provide support and friendship. Either way, however, the newly transitioned person is left in genderland limbo. It is little wonder that after a year or two most transsexuals move on to new positions where their transsexual status is unknown.

A concern carried over from the era before gay openness, was how anyone who worked in positions that required a Security Clearance could transition and be able to maintain their job. That concern had to do with the potential of a gay person being blackmailed by someone interested in gaining access to confidential information. The forced openness of transition especially if done "on the job", precludes any possibility of being subject to blackmail. Nonetheless, I have been interviewed several times (twice, at five-year intervals for one client) by agents of the federal government wanting my opinion on whether or not the individual I was treating was of sound mind and if I would trust that person with national secrets. To my surprise, none of these interviews lasted more than 10 or 15 minutes, the questions were straightforward and none of my clients were denied their security clearance.

Establishing a Healthy Transsexual Identity

Adults who have been struggling with gender issues virtually all their lives enter therapy holding a love-hate relationship with their transsexuality. The "love" part of the equation is tied to a sense of their true gendered self that few would willingly give up even if given the choice. The "hate" part is also understandable, given the burden that having such an overarching issue to contend with on a daily basis has meant for them all their lives. A "Why me?" attitude is not unusual.

The burden has a way of growing heavier and heavier over the years of trying to overcome their insatiable cross-gender role desires. When one adds extreme religious and societal pressures to not even think about wanting to be the other sex to the mix, most enter therapy having a very low sense of self-esteem and little to no expectation of being able to make their lives work.

The real life experience period is an opportunity to try to repair the damage the years of self-destruction have wrought. This is a time for a reawakening as new hormones start kicking in, long-desired physical changes bring dreams into reality and actual person-to-person experiences enliven an otherwise grim existence. These new, positive feelings provide an opportunity for the therapist and others to encourage the individual to move beyond their negative thoughts about their transsexual identity.

People who transition from one gender role to another should be tutored on the realities of what they can expect as a standard part of their therapy. Most gender specialist who are themselves not transsexual—though they mean well and are trying their best—have only the binary understanding of the world they live in to try to shoehorn their client into. They then work to help the individual go from feeling "abnormal" in their assigned sex to "normal" in the desired sex. Since being normal is a highly prized state, most people enter transition thinking that they will leave their abnormal gender-variant condition behind and come out of it a normal healthy man or woman. Unfortunately, that is not what happens. Too many gender therapists use the terms male and female as absolutes in their work. Many gender therapists really seem to believe you can turn a man into a woman and vice versa. To encourage that thinking in their gender clients is doing them a disservice and should be avoided.

The SOC advises that the period of Real Life Experience be at least one year (some countries with a national health care system require a minimum of two years) prior to the individual becoming eligible for sex reassignment surgery. Although some therapists and members of the online gender community consider this period too long and an unreasonable imposition on the transitioning individual, my own experience with more than 500 patients has convinced me that a period of at least 18 months would be more appropriate for all that the transitioning individual needs to accomplish for a successful post-op life.

As we have seen, the physical changes alone require time to make a significant difference in an individual's appearance. It also takes time and energy for the individual to convince both themselves and those close to them that this is the right thing for them to be doing. It is a daunting task not unlike trying to build a house from the inside out. On the other hand, a Real Life Experience of two years may be too long. I have often likened this pre-op or RLE period to fruit ripening on a tree. Picked too early and the fruit is edible but not at its best. Picked too late and the fruit may have started to spoil.

Chapter 8
False Starts, Flip Flops and Regrets

In most cases, gender-role transition takes a linear course. Over time the individuals involved usually find a way to successfully and happily achieve their stated goals. However for some the course is irregular. This chapter looks at aspects of transition that vary from the norm. There are people who start transition only to terminate early on. Others complete transitions, experience their new lives negatively and decide to return to living in their assigned birth sex. Then there are people who have out and out regrets for having transitioned.

There are only a limited number of empirical studies looking specifically at individuals who have not benefited from gender-role transition. Perhaps because the numbers involved appear to make up such a minor percentage of the total number of patients treated, they are not thought of as being of significance. Another possible factor is that these individuals tend to feel that they have been ill served by providers and drop out of sight.

As we have seen, gender role transition starts with a period of psychotherapy, moves on to hormone replacement therapy, life tested by undergoing a prolonged Real Life Experience and then culminates with sex reassignment surgery. Along the course the transitioning individual is professionally guided by a psychotherapist and served by medical professionals well versed in applicable endocrinological and surgical procedures. However, all outcomes are not necessarily what one would hope they would be.

There are two prominent empirical studies that deal specifically with "regrets". The earliest is Friedemann Pfäfflin's 1992 study[1] wherein he concluded that less than satisfactory outcomes were largely due to "poor differential diagnosis, failure to carry out the real-life- test, and poor surgical results". Pfäfflin also noted that in addition to the above, in three cases he observed, maltreatment by the provider played a major role.

In a second study, A. J. Kuiper and P. T. Cohen-Kettenis [2] designed their study to see why some postoperative transsexuals choose to reverse their transition. They interviewed 10 post-op individuals in the Netherlands who reported having feelings of regret or "whose overt behavior indicated a degree of non-successful post-operative functioning, possibly associated with regret." They concluded that "the majority of this group had a (very) late start of cross-dressing and serious psychological problems, which seem to be a result of their gender dysphoria, before requesting SRS."

As we have learned, in male-to-female individuals this course can take as long as two to three years to complete. In many cases even longer. The course, although simply defined, is designed to teach the transperson how to engage their inner questioning of gender identity as well as how to manipulate society into providing space for its manifestations in daily life. By its very nature, the course is a minefield rife with potential failure points. Female-to-male individuals navigating this course can do so in less time then male-to-females but the pitfalls along their course can be almost as treacherous. Furthermore, we are leaning that potential failure does not end with sex reassignment surgery. Problems do arise in some individual's lives years after surgery. We will learn more about that later. In some ways the answer to the question, "Do you have any regrets?" may be different depending on when it is asked.

Much of what has been reported in the literature regarding post-operative regrets, I have seen in my practice as well. It is important, however, that we do not conflate "regret" with "I'm sorry I transitioned because it turns out that I really was a man (woman) after all." It is not that simple, and why some people change their mind about what has transpired is complex. For example, some are disappointed that they experienced bad surgical results and unable to enjoy or even participate in sex any more. There are a subset of regretters who complain that they were not properly informed of what to expect and are now disappointed that they have not become a "real woman" or a "real man." Others complain of having been cleared for surgery too soon, leaving them psychologically disoriented. Another subset of disappointed individuals have mixed feelings: although they are glad to no longer be gender dysphoric, they find that the social costs (partial or full loss of family acceptance, employment discrimination, no longer able to establish or maintain an intimate relationship) is more than they can bear. Others, especially male-to-female individuals who do not pass easily, may learn from continual social rejection and discrimination that they are viewed as misfit males rather than as women, not really fitting in anywhere. It is very hard to live in a world where some people consider you, by your very existence, an unreasonable imposition on their sensibilities and feel free to let you know about it.

Disappointment and the feeling that it would be best not to continue can come anywhere along the process. There are those who regret having started the process and drop out early. Others make it well into transition only to find that outside pressure to stop is more than they can handle. Their regrets have to do with having to stop far short of resolution. Still others complete the process all the way to sex reassignment surgery only to find post-op life so compromised that they would be quick to wish things had been different.

Early Regrets/False Starts.

The further one travels along the road to self-realization through gender-role transition, the harder it gets both physically and psychologically to turn back. To compound the issue, transition does not occur in a vacuum. The further one gets into transition the more one experiences pressure from significant others, friends and even employers to abandon the quest. Rarely is this not the case. There seems to be a big difference between telling someone close to you that you are in therapy for gender identity disorder and considering gender-role transition and showing them the reality of the situation as the transition progresses. This seems to be a bigger problem for genetic men transitioning to the female gender role than it is for genetic women transitioning to the male gender role.

Transition is a long, expensive and painful ordeal no matter what way the transitioning individual is going. If the individual is married and has children, family resources will need to be shifted away from the family and toward the transitioning member. This may be difficult for individuals who may have been the primary breadwinner and who have longstanding beliefs that it is their obligation to support their family but are now asking the family to go with less so that they can do something the rest of the family may find abhorrent. As I pointed out before, transition, by definition, is a completely selfish act. Going on with the imperative of gender-role transition at this seemingly too high a price can force the client to decide not to continue. Those who have the financial means and the support of their families have a better chance at overcoming the difficulties and making it through the process.

Mid transition land mines:

Financial setbacks such as the loss of a job, physical setbacks that limit hormone intake and emotional setbacks such as having a new romantic interest come into one's life are three of the most prevalent obstacles a mid-transitioner might need to navigate.

In the mid-1990s I was treating an MTF individual in her mid-30s. After having been on estrogen hormone replacement therapy for two year she was ready to start living full-time in the female gender role. Letters were written to her employer and all official documentation, including driver's license, Social Security and a legal name change (Roger to Rachel) were accomplished in a matter of weeks. The only change her employer—a large department store—made was to transfer Rachel from the television/large appliance department she had been working in to the housewares department. There she quickly made friends with a heterosexual cisgendered woman. In time the friendship evolved into a serious relationship. A few months later, on a routine visit to my office, Rachel showed up with her girlfriend and notified me that they had fallen in love and were going to get married. The kicker came when Rachel told me that as part of that agreement, she was succumbing to her fiancé's demands that she go back to living as a man. I told her that was, of course, her choice to make and wished her and her girlfriend the best. Rachel dropped out of therapy, stopped taking estrogen and with the help of her still-intact testes, naturally re-masculinized, changed back the identity markers on her official papers and became Roger once again.

About a year later, I got a phone call from Roger requesting an appointment. He complained that his need to cross-dress had returned as strong as ever and that it was "freaking" his wife out. I set up an appointment with him that he cancelled the day before we were to meet. This happened four more times over the next three years. His explanation was the same each time; the two would argue over his threatening to resume transition and he would relent at the last minute, cancel his appointment with me and remain in the male gender role to keep the marriage intact. Eventually, Roger not only made the appointment, he attended it. He informed me that he and his wife had filed for divorce and he was finally ready to resume his transition. This time Rachel completed it and is to my knowledge still living in the female gender role.

There are other forces at work against the mid-term transitioner. A partial list includes fear of being alone, unloved and socially isolated. Being neither male nor female in behavior and appearance can and often does test even long-held relationships to the breaking point. Under the guise of we-are-doing-this-for-your-own-good "tough love motivation," mid-term transitioners may find themselves disinvited to family functions on holidays and even to weddings of their own children. Another pressure point is having the mid-transitioner be forbidden by court to see their minor-aged children unless he or she presents in their birth-assigned gender role. Even then it is not unusual for the court to require the visitations be supervised by a court appointed social worker.

Dropping out of treatment and going back to suffering one's gender issues in private can seem to be the easiest answer. It is, however, often emotionally heartbreaking, leaving the individual in deep despair. The hope and promise of finally resolving their gender issues, held only months earlier, become dashed. Going from a promising high to a hopeless low in a matter of weeks or months can be deadly. It is at this mid-transition time that thoughts of suicide can become their strongest.

Case History of Transitioning, Detransitioning and Retransitioning.

One of the most interesting people I have worked with involved a genetic male who scheduled an intake appointment using the name Stephanie. Over the phone, Stephanie told me that she had transitioned to the female gender role—complete with SRS—16 years earlier and was now in a live-in relationship with a cisgendered female. The couple were experiencing relational problems and she wanted to see me about resolving them. What she didn't tell me prior to the appointment was that she had been taking testosterone for the previous year, largely by request from her partner, ostensibly to increase her libido. She was obtaining the hormone through the same doctor who had provided her original estrogen regimen. Beyond just increasing her libido, it was obvious that her current dose of testosterone was sufficiently high as to remasculinize her.

Along with her relationship concerns, Stephanie also complained about discrimination at work. She worked for a state agency at a position she had been hired into shortly after her original transition. She had worked there quietly for about 13 years, and according to her, everyone assumed she was a genetic female. Pictures she shared of herself taken shortly after her surgery revealed an attractive but otherwise unremarkable woman. Her current problems started when the woman that she eventually became intimately involved with was moved into her cubicle as a workmate. The two started to date and it was on her new friend's urging that Stephanie started to remasculinize. Over time Stephanie began to look more and more like a man but offered no explanation to her employer or anyone else at work. Stephanie slowly but surely became marginalized in her interaction with her co-workers and felt she was being discriminated against. To help ease the discrimination, I advised her to "transition on the job"—that is, admit to being a female-to-male transsexual, but that didn't seemed right to her and she refused to do so.

The relationship issue was the hardest to deal with. Stephanie's partner was of Asian heritage and did not think of herself as a lesbian. To make matters worse, the partner was physically abusive and had multiple sexual encounters outside the relationship that she was open about. Her excuse for her extra-relational affairs was that she needed to be with a man who had a penis. There was another major issue that bothered Stephanie's partner: Stephanie's frequent need to "cross dress." The uncontrollable need to dress in as feminine a way as possible returned more strongly than ever for Stephanie when she began taking testosterone. The disgust over the cross-dressing finally drove the partner to leave the relationship. With her gone, Stephanie stopped giving herself shots of testosterone and resumed her estrogen intake. Her body refeminized easily and she returned to her pre-FTM life. At this point Stephanie terminated therapy and I lost contact with her.

Conclusion:

Even though gender expression deprivation anxiety seems common to clients with gender issues, each person has their own experience of that anxiety. Furthermore, each person lives in a discrete social world from which he or she must deal with the problem. Family, career and social obligations can have a great influence on the process. Even if transitioning was the right thing to do at the time, that choice may not stand up over the rest of one's life.

Patients and practitioners alike need to keep in mind that gender issues are physically complex and loaded with social implications that change from day to day. There are far too many outside influences on the transitioning individual to expect that every case will go smoothly. It is very important to keep in mind that regrets, regressions and even restarts routinely happen in the best of gender centered practices and are part of the the therapeutic process. Providing patience and therapeutic support for the client is all a therapist can do in such situations.

Chapter 9

Life in the New Gender Role

Most books and articles on this subject end with the individual "completing" transition by having a "sex change" operation. As interesting as the transition period seems to be to everyone, including the individuals aspiring to or in the midst of transitioning, it turns out to be only an intermediate period in the individual's life. This chapter describes life immediately after surgery and the repercussions transition has on the rest of the person's life.

Genital Surgery...A true turning point.

Even though genital surgery comes after a prolonged period known as the Real Life Experience (RLE) designed to help both the transperson and those around them to adjust to the realities of the developing gender-role transition, genital surgery changes things dramatically. In fact that is an understatement. Far from being merely cosmetic surgery, as some, especially insurance companies like to think, life takes a turn where in the individual is now permanently consigned to a parallel universe: not male, not female but a biological and sociological combination of both. A true existential point of no return has been breached.

The first thing most post-ops notice, no matter what direction they have transitioned, is that their bodies automatically begin to own their new habitat. Transition moves from being a mental exercise under careful control to being an involuntary (hopefully welcoming) transformation, very much like adolescence. You simply have to hang on for the ride. This is especially true for transwomen who no longer have their testes to influence hormonal levels. For them the default condition is now female. Although it is advisable to continue to take low doses of estrogen to maintain bone health, there is no longer a biological imperative to take massive doses of estrogen to maintain the new female habitat. Only the exogenous taking of testosterone and a bilateral mastectomy could undo the feminization they have experienced should they decide to de-transition.

The situation is different for transmen. Transmen have a choice of either keeping their reproductive organs in place or having them removed. For those who choose to have their reproductive organs removed, their habitat will remain male even if they should choose to stop taking exogenous androgens. However, given the expense and major surgical interventions involved, many transmen choose to limit surgical procedures to bilateral mastectomy and clitoral release procedures (metoidialplasty) retaining their reproductive ability. Doing so, however, requires the continued intake of exogenous androgens at a significantly high dose to maintain ovarian stasis. These individuals retain the ability to return to full fertile capability should they discontinue taking testosterone. (More about post-op transmen pregnancy below). Even so their body seems to know it is now male and responds at a male level of sexual energy and physical ability.

Once the existential point of no return is breached, relationships change dramatically. It is not uncommon to find that in the days and moments before surgery that some of the individual's closest friends and family members have been harboring quiet hopes that the transperson would not really go through with it. Threats such as "You will never see your children again" or " I will disown you" from spouses, parents and siblings have been known to occur as the individual literally enters the hospital. It is the beginning of months or even years of being accused of killing (who you used to be) by some of the people they hoped would have a better understanding of what is occurring.

I am sad to report that in far too many cases these are not idle threats. Although the record is improving, there are still post-op transsexuals who transitioned in the 1970s and 1980s who report being so thoroughly disowned by their families, that they have not seen their children since they transitioned. There are reported cases of some individuals being disinvited to their now grown children's weddings or not even being told of the possible existence of grand children. Others, more fortuitous, report a renewed but awkward association with their now adult children and in some cases, grandchildren.

Starting in the 1990s, as transsexualism became to be seen less as a moral deviancy where by the individual was rendered unfit for parenthood and seen more as a medical problem, we began to see an effort by families to maintain a level of commodity between child and transparent. It was common thinking prior to the 1990's that children had to be protected from the "deviant transparent" less the vulnerable child be negatively influenced (i.e. become gay or transsexual). As time has progressed, and none of the fears of transsexualism being a contagion and unless there are mitigating circumstances, the visitation rights of post-op transsexuals have been greatly liberalized and more and more transsexuals are being seen as loving and caring parents.

Post-op Blues

Some transsexuals experience a let down after their surgery. Every SRS comes after years and years of waiting and preparing. For some the wait has been for decades. Then the big day comes, the surgery is performed in about four hours and the individual is left with a new body that hurts and needs attention in a way the individual has never had to pay attention to before. This is especially true for transwomen who, before they are even healed must start to dilate their new vagina four times a day, learn to pee differently and incorporate a new vaginal hygiene into her daily routine. And that is if all went well with the surgery.

Who Am I Now?

Prior to treatment, transsexuals have a strong tendency to identify with their gender dysphoria. It has been with them so long, it has in their minds, become a major part of their identity and who they are. However, once treatment starts, the administration of cross-sex hormones, the resulting physical body changes and the new and exciting social interactions start to open up spaces in daily existence that they previously only could dream of. It is an exciting period full of promise but alas with a down side. Life for no one is ever all roses and lollypops, that is even more significant for transsexuals where the completion of a life's dream of transitioning and undergoing SRS forces the individual to confront a post-op period where they face several new existential challenges.

For the newly post-op the rules of social engagement are all now very different. Combine that with different genitalia, a head full of cross-sex hormones that they are still learning how to manage, the possibility of an uncertain support network, and now with the dream completed we have a situation that is ripe for a psychological let down.

The quality of life immediately after SRS is greatly dependent on the individual's support network and realistic expectations going into surgery. Having a loving family to return to that can help with the physical healing and embrace the new reality in a loving manner is optimal. Unfortunately, the situation is the opposite of what many post-ops face on return from the hospital. For many it is a return to a lonely apartment with only the relentless schedule of dilations to mark the time. Others, a little more fortunate, can count on other TS friends to keep them company and help with the nursing duties. Once the fog of the anesthesia and pain drugs have worn off, the realization and finality of what the surgery has rendered can be very sobering.

Eventually the newly post-op must get up out of the recovery bed and go back into the world. For those individuals who were working at the time of their surgery, that may or may not be a welcoming event. There always seems to be those fellow workers who disapprove of gender-role transition in general and most decidedly against having one's delicate body parts surgically altered. People can be very hard on post-ops, for example, purposely using the wrong pronouns or worse yet, referring to him or her as an "it". If there are other forces working to diminish the post-op's valuation of themselves and if there is not a good support network to temper the issue, this situation can and has led to suicide.

There are other issues that individuals face upon recently completed surgery. One of the most important is securing gender friendly medical help. Some individuals may be able to stay with the doctor that monitored their hormone regimen. But even then, unless that doctor is involved in transsexual medicine on a daily bases, he or she may have to wing their way in figuring out something as basic as are they treating a male body or a female body. This is especially problematic with transmen who retain their reproductive organs. To make matters worse, there is little or no evidence based medical science in the literature addressing the long term effects the administration of cross-sex hormones has on the body. Fortunately anecdotal evidence collected over the last three to four decades of post-op observation, has shown that there is little or no heightened risk of common health issues in post-op individuals. Drs. Jamie Feldman and Joshua Safer have written the most comprehensive review to date of this subject in the International Journal of Transgenderism[1,2].

Fertility and Reproductive Issues

Transwoman's reproductive options are limited to banking sperm before (or shortly thereafter) starting to take feminizing hormones and ends abruptly with SRS or an orchiectomy. For transmen who opt to keep their reproductive organs, the options for having the ability to reproduce remain. For these individuals, all they need do is go off of testosterone long enough to regain their menstrual cycle, become pregnant with the aid of artificial insemination and carry their own child. The transwoman on the other hand, if she had banked sperm, will either have to find a cisgendered female partner or hire a surrogate to carry her child.

There is a short window in which transwomen can stop taking estrogens and reverse the process long enough for adequate spermatogenesis to resume but exactly how long is unknown. It is highly recommended that the individual bank sperm prior to starting hormone therapy if having a child in the future is at all a concern.

Far more spectacularly, we have seen several occasions where in transmen have decided to get pregnant years after they have reestablished their lives as men. Although there are several reported similar pregnancies over the last 25 years, the most famous transman to put a face on this phenomenon is Thomas Beatie of Oregon. As of this writing he was pregnant with his third child [3]. Mr. Beatie is known to have in a prior period taken sufficiently large doses of testosterone to fully masculinize. He followed with a bilateral mastectomy but opted to keep his female reproductive organs. Getting pregnant was a simple matter of going off testosterone long enough for his menses to resume and have himself artificially inseminated by a male donor.

Stealth

The word stealth has taken on a new significance in the post-op transsexual world. Simply put, the term is used to represent the level of openness a transsexual is willing to share regarding their transsexual status. As stated earlier all post-op transsexuals live in a parallel world where it is prudent to keep information regarding one's transsexual status limited to only those who need to know. There are some male-to-female individuals (usually the very young ones) so deeply stealth that they have gotten married not having told their husbands of their status. The problem here is obvious. One such individual I have worked with lives in constant fear of her husband finding out her genetic male biology. She has no real idea how he will take the news if it should accidently be revealed to him but she feels that she is far too deep into this deception to volunteer the information now.

Most post-op transsexuals learn how to walk the line between being open about their transsexual status and being very closed about it. For everyday matters, with the exception of the most obvious flamboyant person, running stealth is easy and the preferred path of least resistance to go about one's business. Even in situations where the individual takes on a new job or moves to a new community, it is easy and advisable for the individual to keep their transsexual status to themselves. Although all transsexuals live with the fear of being harmed physically for being who they are, there are more reasons other then endangerment to remain unnoticed.

For the most part, transsexuals want to function as authentically as possible in society in the gender role of their choice. Other then self fulfillment, it is the be all and end all for transitioning in the first place. The more authentic the experience feels the more satisfying.

In a twist of irony, the term transsexual has come to embody the very living definition of 'fake'. (i.e. Not a "real man". Not a "real woman"). This is a very bitter pill for most transsexuals to swallow but it has become the as-lived condition of their lives.

I think most transsexuals would prefer to be 100 percent stealth but even if this position was possible there would be severe drawbacks. Human beings, being the social animal that they are, need to talk about their lives with others. Transsexuals are no different in that regard. Perhaps because of the degree of divergence from the norm, that need is greater than that of the average person. There are an inherent subset of issues in a transsexual's life that only another transsexual can relate directly to. None-the-less there are those post-op transsexuals that avoid other transsexuals lest they be outed themselves. By stint of their own transphobia, they end up turning what is already an inherently lonely existence into a very difficult living situation.

The answer, of course, is to be stealth when you need to be and open when you are in a safe environment. The most successful transsexuals I know lead double lives. They have transsexual friends that they meet with and speak to regularly and they have non-transsexual friends who may or may not know of their status. It is with this latter group that they can share non-transsexual aspect of their lives. Keeping tract of who is who and who knows what about one's transsexual status takes skill and alert organization. Ironically this arrangement tends to emphasize the transsexual dilemma but given the alternative it seems to be the best solution.

Sexuality/Relationships

Most people are considered either heterosexual, homosexual or bisexual and that is fixed over the life span. However, when it comes to sexuality, transsexuals find yet again that they are in a parallel universe. Empirical studies comparing pre and post-op sexual attraction preferences are limited. However, in one telling study of 232 male-to-female transsexuals, sexologist, Anne Lawrence[4] found that prior to surgery 54% of the participants said that they were attracted to women and 9% said they were attracted to men. Post-operatively, Lawrence reports, that figure changed to 25% and 34% respectively. Lawrence does not speculate on what may have caused the change but I would propose a possible answer to be men believing that they are approaching cisgendered females, can and do choose to make sexual advancements on transwomen. How that is experienced by the transwoman can very from person to person but at the minimum, most transwomen report experiencing the advancement as a strong confirmation of their femaleness. An experience not unlike the strong feelings of femininity they experienced when they first cross dressed. It can be overwhelmingly positive and a life changer.

Another interesting bit of information that came from the same study, Lawrence found that 85% of the participants reported experiencing orgasm, at least occasionally, after SRS. This figure attest to the improving surgical skills of SRS surgeons.

Despite the positive feelings and the newness of what appears to be a straight foreword male-female sexually charged interchange, it is common in my practice to see that most transwomen find themselves in female/female relationships post-operatively. That can be either with a cisgendered female or another transwoman.

If the transwoman was married at the time of transition and there was no break up of the marriage, a newly defined cisgendered female/transwoman relationship is formed. Surprisingly most wives of transwoman report this arrangement to be far superior to the standard male/female arrangement they once had. It makes sense when one keeps in mind that it is common for these couples to be coming off a prolonged pre-transition period steeped in all the negative angst of a gender dysphoric "husband". Beyond the relief factor, most wives seem to genuinely enjoy having a partner that has all the time and inclination to share in her feminine world.

Another common arrangement is when a transwoman couples with a transman. This looks very much like a heterosexual relationship to the outside world. There is a strong tendency for such couples to raise children either by adoption or the transman carrying one or more children using donated sperm from a friend or relative of the transwoman.

Other common partner arrangements include transman/gay man, transman/straight woman. There are other arrangements as well. However, the important point here is that in the parallel universe of the transsexual, love is freed of cultural limitations.

Transsexual's relationship with one's self

When one talks to those individuals who transitioned decades ago, we learn that the experience goes something like this. They report that they have gone from not quite male to not quiet female. Or not quiet female to not quite male...as the case may be. Although one's sex doesn't change, once cross-sex hormones and surgery are introduced to a mind and body hungering for relief, something significant and profound does occur. What that is, is still a mystery. What has become clear, however, is that post-op transsexual people live permanently in a world of their own. It is a world outside the binary with no idiosyncratic language to explain itself, no image to show its shape, and no legal standing other then what they are able to beg, borrow or steal from those who's world they wished they might some day inhabit but have now left forever out of reach.

On the plus side, many post-op transsexuals speak of experiencing an intellectual and emotional break through with the administration of cross-sex hormones. Surgery seems to cement the new abilities into place. It is as if the administration of cross-sex hormones mature neurons in the brain that have hungered over the life time to be fed. The world opens up to reveal new avenues for creativity; artist, musicians and writers seem the most effected. Post-op people often speak of understanding their own and other people's humanness in a way that had previously eluded them.

I know in my case, after wondering in the wilderness for 40 years, now free from the tyranny of gender dysphoria, I experienced a new expanded sense of time and place. In the span of eight years I completed the requirements for my doctorate, moved to the San Francisco Bay area where I established relationship with a major gallery and completed enough work to participate in several group and one woman painting shows. At the same time I completed my Psychological Assistantship, got my license to practice psychology and went into private practice. I also became interested in my local community governing organization and served as President of our local village association for three years.

Over time, the hormones and continuous interaction with society pushes the post-op transsexual closer and closer to the end of the spectrum they have been seeking. As good as that may sound, transsexuals live with the knowledge that they will never get to where most non-transsexual people start in their understanding of who they are as men or women. Every transsexual is born into and eventually dies in a gendered-self world of their own making.

As off-putting as telling transsexuals entering treatment that they will be going from not quite male to not quite female or vice versa, may sound, very few people have hesitated in going forward based on that information alone. I make sure they understand that they will be going from living a forced gender role they would not have chosen for themselves to a voluntary gender role they may find far more comfortable. The key is the term "gender role" and understanding the sociological limits that accompany the term.

Chapter 10

Summary and Final Thoughts

After a brief summation of the physical and mental health complexities all transsexuals live with, this chapter presents some final thoughts. Experience has shown that there is an enduring dark side to being transsexual and there is an equally enduring bright side. The dark side is characterized by external, societal factors steeped in ignorance, misunderstanding and outright homophobic prejudice forcing the post-operative transsexual to establish a partial to full underground existence. The bright side is characterized by relief from debilitating gender dysphoria, a mix of delightful surprise and muted disappointment upon assimilating the person long known to exist inside. In addition, free of gender expression deprivation anxiety, post-op people, despite the burden if secrecy, tend to melt invisibly back into society, come to terms with their losses over time with many going on to flourish.

As we have shown, gender identity, our innate sense of knowing we are male, female or somewhere in between, is hard wired into our brains by a biological gendering process that occurs before we are born. We have also seen that nature's developmental sexing of the body as male, female or a combination of both while still in utero is a complicated process. Given the complexity of sexing the body and genderizing the brain, it is amazing that things go well as often as they do. Most genetic males have normal-looking genitalia, understand themselves to be male and have no quarrel with that. In like manner, genetic females have normal-looking genitalia and accept their female gender status. In most cases, mind and body are congruent. In recognition of this norm, virtually all societies have decided they should organize themselves on the bases of a binary sexing system.

There are, however, a significant number of individuals who do not survive the embryonic development process with their genitalia looking normal and/or with their sex and their gender identity nicely congruent. These are the individuals who from an early age embark on a life-or-death struggle for peace of mind and a place in society. Fortunately for these individuals, medical science has come up with a solution—of sorts. It is now possible to administer medication (exogenous cross-sex hormones) capable of fulfilling the brain's inexhaustible need for hormones their bodies are otherwise incapable of providing. The reason medication alone is not a complete answer, however, is that the process, invariably started in adulthood, is only partially effective in undoing a lifetime of experiencing the "wrong" hormones, the wrong socialization and the effects of society's binary sexing system. This is the dilemma individuals, who I have been calling transsexuals, face. We now know that the only way out of the dilemma is for transsexuals to face the reality of their biological condition and seek treatment.

We tend to think of birth and death as being the two most important events in life. I would argue there is yet a third— gender-role transition. Birth just happens to us. We have nothing to say about it. We may have a little more to say about the circumstances surrounding our death, but even death, more often than not (suicide being the obvious exception) usually occurs without our cooperation. Gender-role transition, on the other hand, requires consensual cooperation, generally occurs in the mid-process of living and permanently redefines a critical, biological element in what it means to be a human being.

Perhaps the biological redefinition is due to the cross-sex hormones having enriched parts of the brain that under previous circumstances had been starved. Or perhaps it is the experience of surviving the social mine field that all transsexuals must traverse to get by from day to day. Then again, perhaps it is the surgeries to modify genitalia and secondary sex characteristics that make the biological redefinition complete. Whatever the reason, transsexuals come to know early in life that far from simply being "wrong bodied" they are different both physically and psychologically from most other people.

Another factor not commonly understood is that being post-op doesn't mean having completed transition. Life lived in a new gender role requires constant psychological adjustment as the individual ages or enters different social circumstances. It does not take a transitioning individual very long to learn that their transsexual gender status, if possible, is best kept to themselves.

Very little of what is common to the general population and thought of as honest, normal behavior seems to fit in the transsexual's world. Simple things such as having to modify how one speaks of one's childhood, motherhood or fatherhood. Does one lie or generalize the terminology to hide the facts? Most generalize. There are countless times in a transsexual's life when he or she must "fudge" the way they talk about their history. As I noted in the Introduction, I have come to call this linguistic fudging as "transspeak." Here is an example.

I have a transwoman friend (surgery, 1979) who was a Marine helicopter pilot in Viet Nam back in the mid 1970s. She is deservingly proud of her service and enjoys talking about it. The problem, of course, is that most people she shares her military past with seem a bit dumb founded by what they are perceiving as a woman having had such an experience. She addresses their concern by saying she was one of the first women to become a helicopter pilot in the military. In a way she gets special accolades for being a strong pioneering woman but she in fact was just another good male bodied pilot at the time.

Using transspeak is often innocent enough and works for everyday interaction with the public and casual friends. By works I mean it allows the transsexual to be perceived and treated-- gender wise-- according to their own self perception. But how long should one hide one's transsexual status using transspeak when casual friends evolve into good friends? More importantly when does transspeak become a problem when and if a close friendship turns into an intimate relationship? Rather then addressing the issue with their partner, some people, despite their heart felt feeling toward the partner, choose to reluctantly abandon the relationship before addressing the issue. Resigning themselves to a lonely existence. Others summon up the courage to be honest about their past, thereby giving their partner the choice of continuing on with the relationship or breaking up. Either way the transsexual is forced to face the possibility of being rejected for who they are in a way that the general population will never have to.

The Dark Side

If you are a hard-working, honest and law-abiding transsexual, it is very difficult to live in a world that professes to value hard work and honesty only to have those virtues negated based solely on your transsexual status. One of the most prominent examples of this contradiction came in 2007 when the city of Largo, Florida, voted to fire their 16-year city manager, Steven Stanton, when it become known that he was considering transitioning to the female gender role.[1] Public hearings were held in which Steven was loudly called everything from a sexual pervert to an "it" and a dishonest person. Not telling the people who interviewed and hired him for the job 16 years earlier that he had a gender issue was seen as a purposely deceptive act. Steven was dismissed on a 5 to 2 vote based solely on his intentions to transition to the female gender role. Steven's misfortunes did not end there. Eventually fully transitioning to the female gender role, Steven became Susan, experienced the breakup of her family and spent two years and 300 interviews before being hired as a city manager in another city.

In early 2010, when Ms. Amanda Simpson, a male-to-female transsexual, was appointed Senior Technical Advisor to the Commerce Department in President Obama's administration, the media created a brouhaha over her appointment and television comedians made jokes about Ms. Simpson's transsexuality as if she did not qualify to be granted simple human dignity. The media paid little or no attention to Ms. Simpson's extensive abilities as a highly regarded former test pilot who had served as Deputy Director in Advanced Technology Development at Raytheon Missile Systems. Instead, they centered their attention on her appearance and her transsexual status. What are the implications here if not that there is a general belief that being transsexual somehow automatically diminishes one's abilities, makes one morally corrupt or, worse yet, a sad clown to be laughed at?

Far too often transsexualism has been mistakenly equated with homosexuality. Christians, selectively quoting the Bible, often consider transsexualism an affront to God based on the belief that God chooses one's sex at birth and to deny God's will is to deny God himself. Consequently, transsexualism is often sensationalized as a pornographic sexual perversion. That perspective in the hands of someone who feels they have been tricked into having sex with someone who used to be of the other gender or a religious zealot who thinks he needs to rid the world of those he feels has offended God, can mean death and/or disfigurement to the transsexual. Unfortunately, that statement is not an exaggeration. Since 1998, Gwendolyn Ann Smith's website, Remembering Our Dead[1](HYPERLINK "http://www.rememberingourdead.org" http://www.rememberingourdead.org), has been chronicling the number of transgendered individuals worldwide who have suffered violent death directly or indirectly because of who they are. The numbers are staggering and they keep rising each year. The Human Rights Campaign website "How do Transgender People Suffer From Descrimination² writes:

> "Our best estimates indicate that one out of every 1,000 homicides in the U.S. is an anti-transgender hate crime. This estimation is based on data collected by the national organizers of the Transgender Day of Remembrance and the Federal Bureau of Investigation....By this count, they estimate that at least 15 transgender people are killed each year in hate-based attacks, although we believe the number to be higher based on transgender people's common fear of going to the police and widespread misreporting."

Even governments have prosecuted transsexuals. Most of the world's Muslim countries have severe laws against cross-dressing that are strictly enforced. The specter of cross living for a year prior to transition as required for surgery by the WPATH Standards of Care can not even be contemplated in some parts of the world. A lone exception among Muslim countries is Iran, which allows genital reassignment surgery but does so for a perverse reason. Conflating transsexuality with homosexuality, the Iranian Mullahs have decided that because homosexuality is disallowed in the Koran, a "sex change" operation would ensure that any future sexual encounters with someone of the same birth sex would be considered heterosexual.

The Bright Side

Lest one think gender-role transition is all negative, we need to acknowledge the vast majority of individuals who have transitioned and found a path to living successful lives.

Two websites launched by Lynn Conway (HYPERLINK "http://ai.eecs.umich.edu/people/conway/TSsuccesses/TSgallery1 .html" http://ai.eecs.umich.edu/people/conway/TSsuccesses/ TSgallery1.html) and (HYPERLINK "http://ai.eecs.umich.edu/ people/conway/TSsuccesses/TransMen.html" http://ai.eecs.umich .edu/people/conway/TSsuccesses/TransMen.html),   have documented the success of hundreds of transmen and transwomen. This extensive display stands as a testament to all of those individuals who have quietly gone on to live new and productive lives in their preferred gender role.

Post-op life is a mix of delightful surprise and muted disappointment. The "surprise" transwomen routinely report is having become more of a woman than they ever thought possible. A similar phenomenon occurs in transmen. Transmen routinely report having become more of a man than they thought they would become. The problem, however, is that no matter what direction the transition, and despite the terminology generally used in the popular press, one's sex never changes. In the end all transsexuals are genetically either male-bodied women, female-bodied men or genetically intersexed. Nonetheless, the good news is that according to numerous empirical studies,[3-6] life in the new physical paradigm is a major improvement over trying to live a life clinically gender dysphoric.

A last word to providers: Too many medical practitioners still have a binary view of gender. They can envision helping their client to go from male to female as long as they believe the individual can create a successful female appearance. A client recently came to see me after having seen a male therapist who professed being familiar with gender issues. That therapist told the client not to try to transition because he would make an "ugly woman"—a devastating reduction of the client's humanity to one of attractive appearance only.

The goal for therapists working with gender dysphoric individuals should not be limited to helping clients to transition, it should instead be to relieve the client of the chronic gender expression deprivation anxiety associated with gender dysphoria.

That opens up a whole realm of possibilities, ranging from encouraging responsible cross-dressing/cross-living to referring the client to a physician for exploratory doses of cross-sex hormones, and helping the client undergo full gender-role transition, with or without surgery. No one has the right to determine what gender role anyone else should live their life in. The therapist's goal should simply be to help their clients live happy and productive lives rather than make gender-role decisions for them. We know with certainty that taking cross-sex hormones can be a very effective psychotropic medication in certain cases. Combine this with sex reassignment surgery and we reach the upper limit of what is physically possible. From there on forward, success in the new gender role depends on the individual's attitude, support network and willingness to accept life as it is and not as one would have wished.

As time has gone on, we have witnessed the still emerging world of the gender-variant become populated with bright new voices. It is hoped that these new voices will eventually sort all of this out. We need our own philosophers, artist and scientist to help define the gender-variant reality. I hope that by having philosophical discussions free of cultural/religious condemnation based on fear and ignorance on one hand and permissive sugar coating of how wonderful finding one's "real self", on the other may sound, we can eventually pass on a better understanding of gender-role transition not only to our clients but the non-transsexual world as well.

ENDNOTES

Introduction

1. H. Benjamin M.D. (1966).The Transsexual Phenomenon. New York: Julian Press.
2. Ibid. 1, p. 34.
3. A. Vitale (1982). History and resolution of sex/gender integration needs in four male-to-female transsexuals. Unpublished dissertation. San Diego, California: Professional School for Humanistic Studies.
4. D. Karasic and J. Drescher (eds.) (2005). Sexual and gender diagnosis in the Diagnostic and Statistical Manual (DSM): a reevaluation. London: Haworth.
5. K. Winters (2008). Gender madness in American psychiatry: essays for the struggle for dignity. Dillon, CO: GID Reform Advocates.
6. A. Vitale (1997). The therapist versus the client: how the conflict started and some thoughts on how to resolve it. In G. E. Israel and D. E. Tarver II (eds.). Transgender care: recommended guidelines, practical information and personal accounts (pp. 251-255). Philadelphia: Temple University Press.

Chapter 1. Brief Description of The Problem

1. J. Money and A. A Ehrhardt (1972). Man and woman, boy and girl: the differentiation and dimorphism of gender identity from concept to maturity. Baltimore: Johns Hopkins Press.
2. J. Money (1995). Gendermaps: social constructionism, feminism, and sexophical history. New York: Continuum.
3. American Psychiatric Association (1994). Diagnostic and Statistical Manual of Mental Disorders IV. 4th ed. Washington, D.C.: American Psychiatric Association.
4. O. Gilbert (1926). Men in women's guise. London: John Lane.
5. E. Gifford (1933). The Cocopa. University of California Publications in American Archaeology and Ethnology, 31 (whole issue).
6. M. Leach (ed.) (1950). Standard dictionary of folklore, mythology and legend. New York: Funk and Wagnall.
7. C. Bulliet (1928). Venus Castina, famous female impersonators celestial and human. New York: Covici Friede.
8. E. De Savitsch (1958). Homosexuality, transvestism, and changes of sex. London: Heinemann.
9. K. D. Döhlera, S. S. Srivastavaa, J. E. Shryneb, B. Jarzaba, A. Siposa, R. A. Gorskib (1984). Differentiation of the sexually dimorphic nucleus in the preoptic area of the rat brain is inhibited by postnatal treatment with an estrogen antagonist. Neuro Endocrinology Vol.38, No. 4, pp 297-301
10. J-N Zhou, J., A. Hofman., L. Gooren, and D. F. Swabb (1995). A sex difference in the brain and its relation to transsexuality. Nature, Vol. 378, pp 68-70.
11. F. M. Kruijver, J-N Zhou, C. W. Pool, M. A. Hofman, L.J.G. Gooren, and D. F. Swaab (2000). Male-to-female transsexuals have female neuron numbers in a limbic

nucleus. J Clin Endo Metab Vol. 85, No. 5, pp 2,034–2,041.

12. R. C. Pillard and J. D. Weinrich (1987). The periodic table model of the gender transpositions: Part I. A theory based on masculinization and defeminization of the brain. J Sex Res, Vol. 23, No. 4, pp 425-454.

13. O. B. Ward (1992). Fetal drug exposure and sexual differentiation of males. In Handbook of behavioral neurobiology, Vol. 11: Sexual differentiation (A. A. Gerall, H. Moltz, and I. L. Ward, eds.). New York: Plenum.

14. K. E. Kudwa, C. Bodo, J-A. Gustafsson, and E. F. Rissman (2005). A previously uncharacterized role for estrogen receptor: defeminization of male brain and behavior. PNAS Vol. 102, No. 4, pp 608-4,612.

15. A. P. Auger, M. J. Tetel, and M. M. McCarthy (2000). Steroid receptor coactivator-1 (SRC-1) mediates the development of sex-specific brain morphology and behavior. PNAS Vol.97, No. 13, pp 7,551-7,555.

16. D. G. Zuloaga, D. A. Puts, C. L. Jordan, and S. M. Breedlove (2008). The role of androgen receptors in the masculinization of brain and behavior: what we've learned from the testicular feminization mutation. Hormones and Behavior Vol. 53, pp 613–626.

17. J. Imparto-McGinley, R. Peterson, T. Gautierm and E. Sturla (1979). Androgens and the evolution of male-gender identity among male pseudohermaphrodites with 5-alpha reductase deficiency genetics. Genital anomalies and intersexuality. Obstetrical & Gynecological Survey. Vol. 34, No. 10, pp 769-773

18. L. Kohlberg (1966). Cognitive-developmental analysis of children's sex-role concepts and attitudes. In The development of differences (E. E. Maccoby, ed.). Stanford, CA: Stanford University Press.

19. N. Eisenberg (ed) (1998). Handbook of child psychology, 5th ed.: Vol 3: Social, emotional, and personality

development. (pp. 933-1016). Hoboken, NJ, US: Wiley & Sons

20. S. Stein (1984). Girls and boys: the limits of non-sexist rearing. London: Chatto and Windus.

21. F. Pfäfflin and A. Junge (1992). Sex reassignment. thirty years of international follow-up studies after sex reassignment surgery: a comprehensive review, 1961-1991. (Roberta B. Jacobson and Alf B. Meier, trans.); IJT (International Journal of Transgenderism)

22. K. Exner and B. Schneritzky (1995). Female-to-male transsexualism: psychological and social follow-up of reassignment surgery in 67 patients. Paper presented at the XIVth International Symposium on Gender Dysphoria, Kloster Irsee, Germany.

23. P. Snaith, M. J. Tarsh, and R. Teid (1993). Sex reassignment surgery: a study of 141 Dutch transsexuals. Brit J Psych Vol. 162, pp 681-685.

24. R. Green and D. Fleming (1990) in Transsexual Surgery Follow-up: Status in the 1990s, Annual Review of Sex Research. Vol.1, pp 163–174.

Chapter 2. What it Means to be Gender Variant

1. American Psychiatric Association (1994). Diagnostic and statistical manual of mental disorders IV. 4th ed. Washington, D.C. : American Psychiatric Association.

2. A. Vitale (1997). Gender dysphoria: treatment limits and options. Notes on gender transition, online at HYPERLINK "http://www.avitale.com/treatmentoptions. htm" http://www.avitale.com/treatmentoptions.htm

3. A. Vitale (2001). Implications of being gender dysphoric: a developmental review. Gender and Psychoanalysis, An Interdisciplinary Journal Vol. 6, No. 2, pp 121-141.

4. K. Zucker and S. Bradley (1995). Gender identity disorder and psychosexual problems in children and adolescents. New York: Guilford.

Chapter 3. Living the Lie

1. S. Brill and R. Pepper The Transgender Child: A Handbook for Families and Professionals, (2008), Cleis Press, San Francisco, CA., 94114

2. D. Ehrensaft, Raising Girlyboys: A Parent's Perspective, HYPERLINK "http://www.informaworld.com/smpp/title%7Edb=all%7 Econtent=t783567629" Studies in Gender and Sexuality, Vol. HYPERLINK "http://www.informaworld.com/smpp/ title%7Edb=all%7Econtent=t783567629%7Etab=issueslis t%7Ebranches=8#v8" 8, Issue HYPERLINK "http://www.informaworld.com/smpp/title%7Edb=all%7 Econtent=g794708771" 3 June 2007, pp 269 - 302

3. H.A. Delemrre-van de Waal, and P.T. Cohen-Kettenis(2006) Clinical Management of gender identity disorder in adolescents: a protocol on psychological and pediatric endocrinology aspects, European Journal of Endocrinology, Vol 155, suppl 1, S131-S137

4. The American Heritage Dictionary, 3rd Edition, (1992), Houghton-Mifflin, Softkey International, Inc.

5. American Psychiatric Association (1994). Diagnostic and Statistical Manual of Mental Disorders IV. 4th ed. Washington, D.C.: American Psychiatric Association.

6. American Psychiatric Association (1987). Diagnostic and Statistical Manual of Mental Disorders III -R 3th ed. Washington, D.C.: American Psychiatric Association.

7. S. J. Bradley, R. Blanchard, S. Coates, R. Green, S. B. Levine, Heino F.L. Meyer-Bahlburg, I. B. Pauly and K. J. Zucker, Interim report of the DSM-IV Subcommittee on

Gender Identity Disorders, Archives of Sexual Behavior, Vol. 20, No. 4, pp 333-343.

Chapter 4. Treatment Limits and Options

1. There were several organizational meetings prior to the formal adoption of the name, The Harry Benjamin International Gender Dysphoria Association (HBIGDA). According to a paper by Professor Aaron Devor read at the 19th Biennial Symposium (April, 2005) of the Harry Benjamin International Gender Dysphoria Association in Bologna, Italy and in private correspondence, Reed Erickson and his Erickson Educational Foundation contributed directly to the formation of HBIGDA by sponsoring the 1st (1969), 2nd (1971), and 3rd (1973) International Symposia on Gender Identity. These constituted the first three symposia in the biannual series that later became known as The HBIGDA Symposia.

2. The Harry Benjamin International Gender Dysphoria Association (2001). Standards of care for gender identity disorders, sixth version. Committee members: Walter Meyer III, M.D. (Chairperson); Walter O. Bockting, Ph.D.; Peggy Cohen-Kettenis, Ph.D.; Eli Coleman, Ph.D.; Domenico DiCeglie, M.D.; Holly Devor, Ph.D.; Louis Gooren, M.D., Ph.D.; J. Joris Hage, M.D.; Sheila Kirk, M.D.; Bram Kuiper, Ph.D.; Donald Laub, M.D.; Anne Lawrence, M.D.; Yvon Menard, M.D.; Stan Monstrey, M.D.; Jude Patton, PA-C; Leah Schaefer, Ed.D.; Alice Webb, D.H.S.; Connie Christine Wheeler, Ph.D.

3. Ibid 2

4. American Psychiatric Association (1994). Diagnostic and statistical manual of mental disorders IV. 4th ed. Washington, D.C.: American Psychiatric Association.

5. American Psychiatric Association (1987), Diagnostic and statistical manual of mental disorders III-R, 3rd ed., rev. Washington, D.C.: American Psychiatric Association.

6. S. J. Bradley, R. Blanchard, S. Coates, R. Green, S. B. Levine, H. F. Meyer-Bahlburg, I. B. Pauly, and K. J. Zucker (1991). Interim report of the DSM-IV subcommittee on gender identity disorders. Arch Sex Behav Vol. 20, No. 4, pp 333-343.

7. F. Pfäfflin and A. Junge (1992). Sex reassignment. Thirty years of international follow-up studies after sex reassignment surgery: a comprehensive review, 1961-1991 (Roberta B. Jacobson and Alf B. Meier, trans.). International Journal of Transgenderism Electronic Books, available at HYPERLINK "http://www.symposion .com/ijt/pfaefflin/1000.htm" http://www.symposion.com/ ijt/pfaefflin/1000.htm" HYPERLINK "http://www .symposion.com/ijt/pfaefflin/1000.htm" http://www. symposion.com/ijt/pfaefflin/1000.htm.

8. H. Benjamin (1966). The transsexual phenomenon. New York: Julian Press.

9. Bibliographic details for these references follows: R. Green and J. Money (eds.) (1969). Transsexualism and sex reassignment. Baltimore: Johns Hopkins Press; T. Kando (1973). Sex change: the achievement of gender identity among feminized transsexuals. Springfield, IL: Charles C. Thomas; R. J. Stoller (1975). Sex and gender. Vol. 2. The transsexual experiment. London: Hogarth; R. J. Koranyi (1980). Transsexuality in the male: the spectrum of gender dysphoria. Springfield, IL: Charles C Thomas; L. M. Lothstein (1983). Female-to-male transsexualism: historical, clinical, and theoretical issues. Boston: Routledge & Kagan Paul; W. A. Walter and M. W. Ross (eds.) (1986). Transsexualism and sex reassignment, Oxford University Press; A. Bolin (1987). In search of Eve: transsexual rites of passage. South Hadley, MA: Bergin & Harvey; H. Devor (1989). Gender

blending: confronting the limits of duality. Bloomington: Indiana University Press; R. Blanchard & B. W. Steiner (eds.) (1990). Clinical management of gender identity disorder in children and adults. Washington, D.C.: American Psychiatric Press; B. Tully (1992). Accounting for transsexualism and transhomosexualtiy: the gender identity careers of over 200 men and women who have petitioned for surgical sex reassignment of their sexual identity. London: Whiting & Birch; D. King (1993). The transvestite and the transsexual: public categories and private identities. Aldershot, UK: Avebury; K. J. Zucker and S. J. Bradley (1995). Gender identity disorder and psychosexual problems in children and adolescents. New York. Guilford.

10. F. Pfäfflin and A. Junge (1992). Sex reassignment. Thirty years of international follow-up studies after sex reassignment surgery: a comprehensive review, 1961-1991 (Roberta B. Jacobson and Alf B. Meier, trans.). IJT (International Journal of Transgenderism) Electronic Books, available at HYPERLINK "http://www .symposion.com/ijt/pfaefflin/1000.htm(1984" http://www .symposion.com/ijt/pfaefflin/1000.htm(1984" HYPERLINK "http://www.symposion.com/ ijt/pfaefflin/1000.htm(1984" http://www.symposion.com/ ijt/pfaefflin/1000.htm(1984).

11. A. Vitale (2009). T-Note 15 —Testosterone toxicity implicated in male-to-female transsexuals? Some thoughts. Available at HYPERLINK "http://www .avitale.com/TNote15Testosterone.htm" http://www .avitale.com/TNote15Testosterone.htm" HYPERLINK "http://www.avitale.com/TNote15Testosterone.htm" http://www.avitale.com/TNote15Testosterone.htm.

Chapter 5. Transition: Therapeutic Interventions

1. H. Benjamin M.D. (1966). The transsexual phenomenon. New York: Julian Press

Chapter 6. Transition --Clinical Interventions

1. World Professional Association for Transgender Health, Standards of care for gender identity disorders, sixth version, (February, 2001) Committee Members: Walter Meyer III, M.D. (Chairperson); Walter O. Bockting, Ph.D.; Peggy Cohen-Kettenis, Ph.D.; Eli Coleman, Ph.D.; Domenico DiCeglie, M.D.; Holly Devor, Ph.D.; Louis Gooren, M.D., Ph.D.; J. Joris Hage, M.D.; Sheila Kirk, M.D.; Bram Kuiper, Ph.D.; Donald Laub, M.D.; Anne Lawrence, M.D.; Yvon Menard, M.D.; Stan Monstrey, M.D.; Jude Patton, PA-C; Leah Schaefer, Ed.D.; Alice Webb, D.H.S.; Connie Christine Wheeler, Ph.D.

2. A. Vitale (2010). New client information, notes on gender-role transition. Available at HYPERLINK "http://www.avitale.com/Newclientinfo.htm" http://www.avitale.com/Newclientinfo.htm" HYPERLINK "http://www.avitale.com/Newclientinfo.htm" http://www.avitale.com/Newclientinfo.htm.

3. J. F. van Kemenade, P. T. Cohen-Kettenis, L. Cohen, and L. J. Gooren (1989). Effects of the pure antiandrogen RU 23.903 (anandron) on sexuality, aggression, and mood in male-to-female transsexuals. Arch Sex Behav Vol. 18, pp 217-228.

4. S. H. van Goozen, P. T. Cohen-Kettenis, L. J. Gooren, N. H. Frijda, and N. E. Van de Poll (1994). Activating effects of androgens on cognitive performance: causal evidence in a group of female-to-male transsexuals. Neuropsychologia Vol. 32, pp 1,153-1,157.

5. S. H. van Goozen, P. T. Cohen-Kettenis, L. J. Gooren, N. H. Frijda, and N. E. Van de Poll (1995). Gender differences in behaviour: activating effects of cross-sex hormones. Psychoneuroendocrinology Vol. 20, pp 343-363.

6. C. Miles, R. Green, G. Sanders, and M. Hines (1998). Estrogen and memory in a transsexual population. Horm Behav Vol. 34, pp 199-208.

7. D Slabbekoorn, S. H. van Goozen, J. Megens, L. J. Gooren, and P. T. Cohen-Kettenis (1999). Activating effects of cross-sex hormones on cognitive functioning: a study of short-term and long-term hormone effects in transsexuals. Psychoneuroendocrinology Vol. 24, No. 4, pp 423-447.

8. P. T. Cohen-Kettenis and L.J.G. Gooren (1993, Modified, 2007). The influence of hormone treatment on psychological functioning of transsexuals. J Psych & Hum Sex HYPERLINK "http://www.informaworld.com/smpp/title%7Edb=all%7Econtent=t904385168%7Etab=issueslist%7Ebranches=5#v5" http://www.informaworld.com/smpp/title%7Edb=all%7Econtent=t904385168%7Etab=issueslist%7Ebranches=5#v5" 5(4):55–67.

9. W. J. Meyer III, A. Webb, C. A. Stuart, J. W. Finkelstein, B. Lawrence, and P. A. Walker (1986). Physical and hormonal evaluation of transsexual patients: a longitudinal study. Arch Sex Behav Vol. 15, No. 2, pp 121-138

10. Ibid. 1, p. 20

11. Ibid. 1, p. 18.

12. Ibid. 1, p. 18.

13. M. Brownstein, (2006), Plastic, reconstructive and gender related surgery. Available at HYPERLINK "http://www.brownsteinmd.com" http://www.brownsteinmd.com" HYPERLINK "http://www.brownsteinmd.com" www.brownsteinmd.com/.

14. M. L. Djordjevic, D. Stanojevic, M. Bizic, et al. (date?) Metoidioplasty as a single stage sex reassignment surgery in female transsexuals: Belgrade experience. J Sexl Med Vol. 6, No. 5, pp 1,306-1,313.
15. M. Bizic, M. Majstorovic, V. Kojovic, D. Stanojevic, G. Korac, B. Stojanovic, and M. Djordjevic (date?). One stage metoidioplasty in female to male transgender patients: the role of genital flaps for urethral reconstruction. Eur Urol Supp Vol. 8, No 8, pp 648.
16. Ibid, 1, p. 21.

Chapter 7. The Reality of the Real Life Experience

1. F. Pfäfflin and A. Junge (1992). Sex reassignment. Thirty years of international follow-up studies after sex reassignment surgery: a comprehensive review, 1961-1991 (Roberta B. Jacobson and Alf B. Meier, trans.). IJT (International Journal of Transgenderism) Electronic Books, available at HYPERLINK "http://www.symposion .com/ijt/pfaefflin/1000.htm(1984" http://www.symposion .com/ijt/pfaefflin/1000.htm(1984" HYPERLINK "http:// www.symposion.com/ijt/pfaefflin/1000.htm(1984" http:// www.symposion.com/ijt/pfaefflin/1000.htm(1984).
2. B. Kuiper and P. Cohen-Kettenis (1988). Sex reassignment surgery: a study of 141 Dutch transsexuals, Archives of Sexual Behavior Vol. 17, No. 5, pp 439-457
3. K. Exner and B. Schneritzky (1995). Female-to-male transsexualism: psychological and social follow-up of reassignment surgery in 67 patients. Paper presented at the XIVth International Symposium on Gender Dysphoria, Kloster Irsee, Germany.
4. Y.L.S, Smith, S. H. van Goozen, A. J. Kuiper, and P. T. Cohen-Kettinis (2005). Sexreassignment: outcomes and

predictors of treatment for adolescent and adult transsexuals, Psychol Med Vol. 35, No. 1, pp 89-99.

5. R. Green (1998). Transsexuals´ children. IJT (International Journal of Transgenderism) Volume 2, Number 4

6. M. J. McConaghy (1979). Gender permanence and the genital basis of gender: stages in the development of constancy of gender identity. Child Devel 50(4):1,223-1,226.

7. Human Rights Campaign Foundation (2008). Corporate equality index 2008. Available at HYPERLINK "http://www.hrc.org/issues/workplace.asp" http://www.hrc.org/issues/workplace.asp" HYPERLINK "http://www.hrc.org/issues/workplace.asp" www.hrc.org/issues/workplace.asp.

Chapter 8. False Starts, Flip Flops and Regrets

1. F. Pfäfflin (1992). Regrets after sex reassignment surgery. J Psych & Human Sex Vol. 5, pp 69-85.

2. A. J. Kuiper and P. T. Cohen-Kettenis (1998). Gender role reversal among postoperative transsexuals. IJT (International Journal of Transgenderism) Vol. 2, No. 3

Chapter 9. Life in the New Gender Role

1. J. Feldman and J. Safer (2009). Hormone therapy in adults: suggested revisions to the sixth version of the standards of care. IJT (International Journal of Transgenderism) Vol. 11, pp 146-182.

2. G. De Cuypere, G. T. Sjoen, R. Beerten, G. Selvaggi, P. De Sutter, P. Hoebeke, S. Monstrey, A. Vansteenwegen, and R. Rubens (2005). Sexual and physical health after sex reassignment surgery. Arch Sex Behav Vol. 34, No. 6, pp 679-690

3. T. Beatie (2010). Pregnant with third child. Available at Advocate.com.
4. A. A. Lawrence (2005). Sexuality before and after male-to-female sex reassignment surgery. Arch Sex Behav, Vol. 34, No. 2, pp 147-166.

Chapter 10. Summary and Final Thoughts

1. G. A. Smith's website, Remembering Our Dead (HYPERLINK "http://www.rememberingourdead.org" http://www.rememberingourdead.org" HYPERLINK "http://www.rememberingourdead.org" http://www.rememberingourdead.org

2. Human Rights Campaign, How do Trangender People Suffer Discrimination? HYPERLINK "http://www.hrc.org/issues/health/1508.htm" http://www.hrc.org/issues/health/1508.htm

3. F. Pfäfflin, A. Junge, (1992) Sex Reassignment. Thirty Years of International Follow-up Studies After Sex Reassignment Surgery: A Comprehensive Review, 1961-1991 (Translated from German into American English by Roberta B. Jacobson and Alf B. Meier); IJT Electronic Books, on-line available at HYPERLINK "http://www.symposion.com/ijt/pfaefflin/1000.htm(1984" http://www.symposion.com/ijt/pfaefflin/1000.htm(1984)

4. K. Exner, & B. Schneritzky, (1995). Female-to-male transsexualism: psychological and social follow-up of reassignment surgery in 67 patients. Paper presented at the XIVth International Symposium on Gender Dysphoria, Kloster Irsee, Germany.

5. P. Snaith, M. J. Tarsh, & R. Teid, (1993). Sex reassignment surgery: A study of 141 Dutch transsexuals. British Journal of Psychiatry, 162, 681-685.

6. R. Green & D. Fleming (1990) in Transsexual Surgery Follow-up in the 1990s. Annual Review of Sex Research.Vol. 1, pp 163-174

www.ingramcontent.com/pod-product-compliance
Lightning Source LLC
Chambersburg PA
CBHW030756180526
45163CB00003B/1054